Engineering Words

Communicating clearly in the workplace

Sharon Burton

Bonni Graham Gonzalez

Second edition

XML PRESS

Engineering Words

Communicating clearly in the workplace

Credits

Cover image	Wind Turbine ©2013 lamoix@flickr. Licensed under a CC BY 2.0 license.[1]
Figure 15.1	Prince Charles ©2017 Mark Jones. Licensed under a CC BY 2.0 license.
Figure 15.1	Ozzy Osbourne ©2010 Kevin Burkett. Licensed under a CC BY-SA 2.0 license.[2]

Disclaimer

Trademarks

XML Press
Denver, Colorado 80230
https://xmlpress.net

Second Edition
978-1-937434-89-2 (print)
978-1-937434-90-8 (ebook)

---

[1] https://creativecommons.org/licenses/by/2.0/
[2] https://creativecommons.org/licenses/by-sa/2.0

Contents

Foreword to the Second Edition vii

Foreword to the First Edition ix

Acknowledgments xi

1. Welcome 1

    You can learn to communicate 2

    Who is this book for? 3

    Terms we use in this book 3

2. Clear Writing Guidelines 7

    Clear communication matters in the business world 8

    The basics 9

    Building sentences and paragraphs 21

3. Writing Good Procedures 27

    What are procedures? 28

    The goal of a procedure 29

    The structure of a good procedure 29

    Using graphics in procedures 34

    Defining task paths 36

    Overview procedures and task-specific procedures 38

    Complex procedures 42

4. Writing Tools 49

    Spell check and grammar check 50

    Terminology managers 51

    AI tools 51

5. The Business Context of Communication 57

    Writing for professors/peer reviewers 57

    Writing for your boss/business peers/the public 58

    So what is the business context? 58

    Communication in the context of the flow of money 67

6. The Workplace Ecosystem 69

    How big is your workplace? 70

    On-premises vs work-from-home 72

    How to be engaged and productive 73

    Leadership in the workplace 84

    Goal-setting and planning 87

7. Résumés and Cover Letters 89

    Cover letters 90

    Résumés 92

    Ways to find a job 98

8. Ethics in Engineering 101
    Your life story: beginnings 101
    Your life story: your present and future 106
    And here's the problem 109
    Who you are is the same person all the time 110
9. Flow of a Project in a Company 113
    Project start 114
    Business requirements 116
    Functional and technical specification phases 117
    Development (and testing) phase 120
    Delivery phase 124
    The secret 127
10. Pitching Ideas 129
    Finance context: how the flow of money fits 129
    Historical/technical context 130
    Market context: the technology adoption curve 132
    Communicating in the market context of the technology adoption curve 136
    Practical application: the business case 137
11. Designing Effective Presentations 145
    Getting started 145
    Some important terms 146
    Slides you should always have 147
    Overall slide deck structure 148
    Text formatting 151
    Colors 153
    Graphs and charts 157
    Spelling and grammar 160
12. Handling Yourself and the Room in Presentations 161
    A day or two before the presentation 162
    Right before the presentation 164
    The presentation 165
13. Cognitive Science 171
    About humans 171
    The physical world and biology 172
    Our vision 173
    The world happens in human brains 173
    Other processes 176
    Learning theory 177
    Humans require explanations 181
14. Constructing Explanations 183
    Explanations and our brains 183
    Design and perception: perceptual explanations 185
    Document to the question: cognitively constructed explanations 191

15. Personas and Scenarios 201
    Why do you care? 201
    You can't develop for yourself or for "everyone" 201
    Personas 206
    Scenarios 209
    Wrong personas and wrong scenarios 212
    What do you do with personas and scenarios? 213
16. Writing Functional Specifications 215
    The nature of a functional specification 216
    Deconstructing functions and features 217
    Writing functions and features 219
17. Testing Your Products 227
    Organizational styles 227
    Why test? 228
    Hierarchy of testing documents 232
    Deconstructing test cases 233
    What makes a good test case? 233
    Structure of test cases 235
A. Metaphors 245
    What is a metaphor? 245
    Not all metaphors are created equal 246
    Personas drive metaphors 247
    Metaphors and examples 247
B. References 249
Index 257

Foreword to the Second Edition

Many years ago, I was working with a group of engineers who were designing what should have been a straightforward algorithm. I struggled to document how the feature worked because the specifications just didn't make sense. One meeting with three engineers later, we discovered that there was a flaw in the algorithm. Had it not been for the need to explain the function in a clear way for users, the problem might not have been noticed until it became an expensive mistake to fix in the product.

Who better to help technical professionals write clearly than two experienced instructors—and highly-skilled communicators—who have come together to share their expertise with a wider audience. Sharon and Bonni have extensive industry experience documenting products and services, as well as teaching clear communication to engineers in educational and corporate settings. They walk the talk; they bring the subject matter to life in an easy-to-read way, making the book an example of what they teach. When you find yourself absorbing the material without struggle, think of how your audiences will do likewise. This means they will concentrate on learning from your expertise in your domain, not revisiting sentences to try to make sense of the topic.

The way we communicate has changed over the years, particularly as neuroscience helps us understand how the brain processes information and as learning theory helps us shape how we communicate with the consumers of our content. The authors have incorporated elements of this in the book as a way of explaining why a particular technique works and how to use it.

Another way we communicate differently is with the corporate push to use Generative Artificial Intelligence to boost efficiency. It may boost efficiency, but not necessarily clarity. The way we use GenAI to communicate affects our communication style once again. Sharon and Bonni cover a wide range of these aspects of writing, which, in turn, will help you avoid some of the pitfalls that can affect your work.

As the saying goes, context is everything. This book covers a wide range of contexts, some of them specific, while others are transferable. Whether you're still in school, a new graduate, or have been in the workforce for decades, you're sure to find something of value. Even though I have been in the content industry for over three decades, I found valuable nuggets of information that I'm incorporating into my work.

Investing in your communication is an investment in your professional future. Being able to apply what you learn in this book is well worth the time spent reading it.

Rahel Anne Bailie is Content Solutions Director with Altuent, an Irish knowledge management consultancy. She has been working with content for several decades, is a published industry author, and teaches in the Content Strategy Master's Program at FH-Joanneum in Graz, Austria.

Foreword to the First Edition

"Technically brilliant, but..."

Many great ideas are lost to terrible communication. Technical competence in your field is of course necessary, but without effective communication skills, your influence and your career prospects are limited.

As an engineer, you know how to attack and solve difficult problems. One of the most difficult problems you'll face is how to explain your ideas to others and convince them that your approach is the best solution. For that, you need communication skills.

But too many of you are trained only on the engineering side. You either couldn't or didn't include communication skills in your education. This gap is a critical defect that you need to address.

When you are building a product, you can look at your work as having three critical requirements. First, the product needs to be safe to use and operate. Second, it needs to solve the problem at hand. Third, it needs to be usable.

If your requirement is to build a bridge that transports cars, a pedestrian bridge, no matter how beautiful, will not be a success. *It doesn't solve the right problem.*

Communication has a similar set of requirements. First, your communication needs to be clear and understandable. Second, it needs to be pertinent to the business requirements. Third, it should be persuasive.

Clear and understandable communication requires you to follow basic rules of grammar and usage. If you make grammatical errors or write overly complex sentences, your audience may have trouble understanding your message. If they don't understand you, your communication effort will fail. Writing well means anticipating and avoiding obstacles for your readers. You don't have to be a literary giant, just a competent writer.

Remember that your audience nearly always includes people with limited language proficiency. Perhaps they are reading your document in their second (or third or fourth) language. Maybe you are discussing a topic in which you are an expert but your audience is not. The average American reads at a 5th- to 7th-grade level. Careful word choice and simple sentence structure help your readers understand you.

Your content should be relevant to your audience, and often, you have to present the information differently for different audiences. Remember that communication is controlled by the recipient, not the author. You write an in-depth explanation of your approach with gruesome technical details, but your target audience is not a technical person. Your document will probably just annoy the reader. Instead, dial back the technical details so that a non-expert can understand what you are trying to say. "They should learn to be more technical" is not an acceptable attitude. You may be right, but unless you meet your audience at their level, your communication will fail.

If your goal is to persuade fellow engineers that your solution is the best option, you probably need an assessment of all the different options and their technical pros and cons. If your goal is to get funding for an initiative, you need to discuss what funding you need and what the business result will be. The technical details don't belong in that content. Think of your communication efforts as being outcome-oriented—start with the result you want and then figure out what information you need to provide to achieve that goal.

There is an entire discipline devoted to persuasive writing—marketing! As an engineer, you may think that marketing is beneath you. You are, my friend, so very, very wrong. Throughout your career, you will need to market yourself for jobs, market your ideas to get funding, market your contributions to get raises and promotions, and so on. The sooner you learn how to write well, the better off you will be in your career.

This book will help you understand that technical brilliance is just the beginning. Read it.

Sarah O'Keefe is the founder and CEO of Scriptorium Publishing, which works at the intersection of content, technology, and publishing. Today, she leads an organization known for expertise in solving business-critical content problems with a special focus on product and technical content.

Acknowledgments

This section isn't important to you, perhaps, but it's very important to us. The year we started working on the first edition of this book was also the year Sharon's husband was diagnosed with a terminal illness. The following year, Bonni's mother was also diagnosed with a terminal illness. We put everything aside for obvious reasons. Bonni's mother died in 2014 and Sharon's husband died in 2016.

Several years later, we thought we were ready to get back to this project when a world-wide pandemic occurred. Everything was put aside again for obvious reasons. Between our full-time jobs, suddenly teaching remotely at our part-time jobs, and managing the pandemic upheaval in our personal lives, it was too much.

When we finally got back to this project, our families were patient and supportive during the time we took away from them to finish this project. Our publisher has been far more than patient with us.

Without the support of our families, our friends, and our students, this project would not have been possible.

Thanks to all of you.

CHAPTER 1

Welcome

The reality of your engineering career is that you will work in a business environment at least one time. You will talk to and write documents for engineers and non-engineers who hold power over whether your ideas are accepted and acted on. You may have brilliant ideas that create new markets, solve problems in new ways, and overall change the world.

But if you can't communicate with other humans, your brilliant ideas don't matter. You must get your ideas out of your head and into the heads of other humans. Other people need to see your ideas and know why they are important and amazing. They can't read your mind, and they are not dumb just because they don't see your vision as clearly as you do. It's up to you to know how to communicate your ideas, to *engineer* your ideas, so that your audience gets it.

Oh, you might say, I'll always be working with other engineers. I don't need to worry about this communication thing because it's all engineering. And you would be wrong. Or at least you're wrong if you want to advance your career and do things other than on-the-ground engineering tasks. At some point, you might want to lead a team or lead multiple teams. You might want to run a company of your own.

Even if you spend your entire career communicating with other engineers, those other engineers probably don't share your exact background or education. They may have very different training than you do. They certainly don't have your brain or your way of thinking. But you must communicate well with them to get the work done.

In your career, you'll write at least:

- Customer stories,
- Emails,
- Functional specs,
- Product development tickets,
- Knowledge base articles,
- Project plans,
- Retrospectives,
- Company chat messages,

- Status reports,
- Support tickets,
- Test cases,
- Use cases,
- and so much more.

Not all these documents will be for other engineers—they're documents that are written in the business environment for your fellow engineers, other co-workers, and leadership. You'll need to communicate ideas and thoughts, so other people can make decisions or at least know what you're doing with your day. Many of these documents are the only record of why things happened the way they did. Your documents are part of the institutional memory of the company. These communications matter to the company.

You can learn to communicate

Communicating with humans is not rocket science and can be learned. This book focuses on teaching you how to communicate in the workplace, but many of the techniques we describe can be used in your personal life as well.

Considerable research in the last 20 years helps us understand how people absorb information, how they process information, how they store and retrieve information. This book uses this research to help you better communicate in the workplace. If you want to go in-depth on any of these topics, we provide references to additional resources in Appendix B, *References* (p. 249).

In this book, we cover:

- what clear writing is and specifically how to do it
- how to use modern writing technology to get your work done faster
- what a business cares about and how to effectively work in one while keeping your ethics
- how a project flows through a company and how to get your ideas approved
- how to present effectively
- how humans think and process information
- how to use all this information to create several common engineering documents

Along the way, we share stories and offer examples to give you context for what you're learning.

Who is this book for?

While the topics contained in this book are appropriate for college-level students, you don't need to be a college student to benefit from the knowledge found here. Practicing engineers who want to improve workplace communication skills can find equal value in this text. The concepts work as well for professionals as they do for college students, and you can easily apply them in the workplace.

College-level engineering students can use this book to learn communication best practices that help them succeed in the workplace, even as an intern. Some concepts can be applied to other writing in some of your engineering and non-engineering classes right now.

People who work in STEM-adjacent positions may also find this book useful. Communication is communication in the workplace. If you're not an engineering student or professional but find yourself communicating with other humans in the workplace, there may be much here for you. Many roles in the technology world are not engineering jobs, but they are still critical to designing, developing, and delivering technology products.

Terms we use in this book

Throughout this text we talk about *minimum viable product*. While this term has many definitions, we mean the least amount of product that meets requirements and can be released on schedule. We're all about delivering products on schedule. Because that's what business cares about.

Often related to the minimum viable product is *technical debt*. During the product development phase, decisions get made to get the product shipped. These decisions can mean reducing the features or taking shortcuts. Eventually, these decisions can make other things hard or impossible to do because of the debt incurred from reducing features or taking those shortcuts. It's not unusual to eventually have one or more releases that do nothing but address this technical debt.

Occasionally, we refer to *enterprise-level products*. These are products that impact or are used by most people in a company. For example, Salesforce is a great example of an enterprise-level product. So is the human resource product your company may use. These products do many things that impact multiple users across a company, from accounting to sales to the shipping department. Creating these products is a considerable undertaking that involves multiple teams working on different areas of the product, often on different release schedules.

We refer to a *waterfall environment*.[1] A waterfall environment completes each phase of a project before the next one is started. For example, the business requirements are completed and signed off before the functional specification phase starts. While it's less common in this century, it remains a valid development methodology for some products. If you're developing hardware, expect to work in a waterfall environment. Rapidly creating and releasing hardware in 2-week sprints (short development periods) is impractical.

We have both worked in *agile environments*[2] multiple times. While there are multiple ways to implement agile, the distinctions are immaterial for the sake of this book. When we talk about agile, we mean product development using a looser, faster method than a waterfall method. For our purposes, agile development typically includes 2-week work sprints, using some sort of burndown task list based on a user story system, and probably some epics, which are a group of related user stories.

> **A note about agile:** At times in this book, we discuss communicating in an agile environment. We may also sound disparaging of the agile environment. Nothing could be less true. We've both worked in agile environments, and they can work brilliantly. However, more often than we like, when a director of development says they work in an agile environment, they mean code like hell and hope for the best as they throw the code at the customers. While code like hell and hope for the best is an all-too-common development environment, it's not an agile environment. This distinction matters.

We also talk about *work product*. A work product is the tangible result of your work, such as a report or presentation or the code you write, that your employer pays you money to create for them. That work product must be created to your employer's specification. That means they can tell you what tools you use and when and how you deliver the work product because that's what they pay you to do for them. Not understanding that your job consists of creating and delivering the work product in the time frame your employer determines can result in unhappiness from your employer. It can also result in you being excused from that employment.

We sometimes refer to *STEM-adjacent jobs*. These are jobs in the technology world near science, technology, engineering, and mathematics (STEM) fields. For example, usability testing and technical marketing are STEM-adjacent. You can do these jobs without a STEM degree, but having one can help. Both authors work in STEM-adjacent fields.

---

[1] https://en.wikipedia.org/wiki/Waterfall_model

[2] https://en.wikipedia.org/wiki/Agile_software_development

Engineering does not happen in a vacuum. Every technology breakthrough has a history of *prior art*—previous discoveries of technology that make the current technology breakthrough possible. For example, if humans had never learned to control fire, we could not have jet engines. Fine kitchen knives trace directly back to the discovery that some rocks are sharper than others and can be made sharper still by breaking them very carefully.

Throughout this book, you will notice that we mix up pronouns. We are tired of seeing engineers referred to only as he or him. We are also aware that in the world as it is in this century, we can be, and should be, more flexible about pronouns. We want to include everyone because everyone that can be an engineer should be an engineer, regardless of their specific pronouns.

Because the machine *does not care*. The only situations where the machine cares is where we've programmed it to.

What do we mean by this? Neither of us has ever had a compiler refuse to work because of our gender. No web page has refused to load because of the color of our skin. No device we have used has broken because of who we love or how we identify. And we don't know anyone else who has had such an experience.

The machine (whether it's software, hardware, a wearable, a car, a clock, a microwave, a smart thermostat, Alexa, Siri, or a screwdriver) only cares whether the engineering is any good. Write good code? Design a good circuit? Build a quality device? Develop a durable and sturdy tool? Create a useful interface? Excellent—the machine will work. Unless programmed by humans otherwise (either inadvertently or on purpose), the machine works regardless of your gender, race, ethnicity, sexual identity, or religion (or lack thereof).

Be like the machine and focus your attention on the quality of the engineering, not on the people creating it.

CHAPTER 2
Clear Writing Guidelines

You've probably taken many writing classes during your education, all of which have been about "teaching you to write." That's not this chapter. We assume you can write—that you can construct coherent sentences in English.

This chapter is about written communication *for business*. This chapter explores a different way of constructing information. Adapting to this new approach may be challenging, at first. That's OK. Most engineers eventually prefer writing this way, as it's clearer and more straightforward.

"Oh," you may think, "I'm an engineer. I don't need to worry about communicating because the genius of my engineering will be obvious to all." And you would be wrong. You must be capable of communicating with many people in the business world. You want your genius ideas to be understood. You can't hope people in the business world will figure it out themselves.

Academic writing is not appropriate in the business world. In academic writing, it can seem you're encouraged to write very long sentences, with many words and many commas, and to just keep rambling on. It can take pages to get to a point, and if the point is vague and unclear, well, that's all the better. You may feel like you got As in school for just this sort of writing.

Academic writing demonstrates you understand the information your teachers tried to teach you. Your work product, if you will, demonstrates your knowledge and comprehension of the material covered in the class. Your instructor is looking for your mastery of the information and assigns you a grade based on how well you demonstrated that understanding in your deliverable.

This type of writing isn't suited for the business world. In the business world, they assume you have the depth of knowledge because they hired you. They don't want you to demonstrate that you know the field so you can get a good grade. They need you to understand business issues and concerns, so business decisions can be made based on the information you present.

In the business world, everyone is busy. No one has time to parse complicated and confusing communication. If it's not clear, the communication is returned to you for clarification (at best) or thrown out and ignored (at worst). Either way, your message is unclear, and it's your fault the noise in your communication overwhelmed the signal—which means it's your problem to solve. You need to engineer your words using the same care you craft your code or develop your devices.

8 Clear Writing Guidelines

Clear communication matters in the business world

In the business world, you need to communicate because that's how the business world works—people communicate with each other. Your boss wants to know what you worked on last week, co-engineers want to know if your work is ready to be integrated into their work, you want HR to grant you a vacation, and so on.

Sometimes this communication happens with people just talking to each other, but asynchronous methods, like writing, support most communication in the business world. You write a weekly report to your boss, you contribute to the spec for the new products you're working on, you want to go on vacation next month, you write a message to a customer in Slack. It's obvious that a company with more than two employees can't function effectively if nothing gets written down.

And that's where clear written communication comes in. If you write it down but don't communicate it clearly, you waste time and no one understands what you're trying to communicate. If no one understands what you're communicating, you won't be at that job very long. If you want to run your own company, no one will understand what you're trying to do. No one will give you funding or buy your product.

Even if you are communicating with other engineers, you need clear communication. Think of the confusion that sometimes happens when you're talking to engineers in your engineering specialty. In the business world, you're working on teams of engineers from other engineering fields, educated in other countries, with different engineering knowledge and expectations. Something as simple as not communicating the units of measurement can be an issue.

If you think that can't possibly happen, consider the following. NASA crashed a Mars Orbiter into the planet. After reviewing what happened, NASA determined one team was using English/Imperial units of measure while other teams were using metric. When the orbiter adjusted its orbit, the different systems moved the orbiter too close to Mars, where it couldn't function properly in the atmosphere. It most likely was damaged and skipped off the atmosphere towards the sun. All the engineering effort to get it built and launched—273 travel days, and $125 million dollars—was wasted, largely because the engineers on the teams did not communicate what they thought was "obvious."[1]

---

[1] "Mars Climate Orbiter Mishap Investigation Board Phase I Report" (NASA 1999)

The basics

This section starts with the basics to refresh your memory. In no way should you think of this as a comprehensive overview of constructing sentences in English or a substitute for intensive English training. If you're uncertain of your ability to write in English (perhaps you're still learning English), seek a class or group to support you.

There are exceptions to almost everything in this discussion, but we're simplifying for our purposes. If we start simple, we can build to more complex.

What's in a sentence?

A sentence is the smallest piece of communication we're going to look at. As you recall from grade school, a sentence consists of a subject, a verb, and then some other stuff. A sentence is a complete thought.

> A new thing that's appeared recently is removing periods at the ends of sentences. Some people think using periods at the end of sentences indicates passive-aggressive behavior.[2] This is not a good thing. Periods are how other people know your sentence is complete—that you finished your thought and started a new thought. Use periods at the ends of your sentences, so people know your thought is complete.

A sentence includes a subject. The subject is the actor of the sentence. The subject is the *who* or the *what* that's doing the thing the sentence is about. For example, our subject is **Bob**. The verb is the action the subject does. For example, our verb is **runs**. The other stuff is in the predicate—everything after the verb. For our example, let's have a prepositional phrase: **to the car**.

This gives us a simple sentence: **Bob runs to the car**.

Technically, we could stop the sentence after the verb: **Bob runs.** In English, we don't need anything else. But most sentences are not two words. Most sentences have quite a bit of information in the predicate.

Research shows that the *subject + verb + predicate* structure is the easiest structure to understand in English. It may also be the easiest in other languages that use this construction. The Plain Language initiatives many governments have adopted recommend this structure as the easiest to help people understand communication.

---

[2] "Are Your Texts Passive-Aggressive?" (Hensel 2020)

Active voice

In technical communication, which is the sort of communication you're doing as an engineer, the most important "voice" is *active voice*. Active voice depends on the subject of the sentence, the actor, to do the action, so the actor should come first in the sentence. An example is: **The writers won the award.**

A common misperception is that active voice adds excitement to your writing and that the best way to add excitement is to add exciting words. Our students often add "action words" to their sentences, thinking that makes the sentence active voice. This is a fundamental misunderstanding of active voice. Review writing texts to clarify if you have the same misperception.

Passive voice hides the actor of the sentence and makes it hard to understand who is doing what. Academic writing often teaches passive voice as a way to remove the author from the writing. Unfortunately, this is not good writing in the business world—we need to know who is doing what. An example of passive voice is: **The award was won by the writers.**

Worse, passive voice makes your reader pause and turn the sentence around in their mind to figure it out. It's not conscious, but this pause impacts comprehension and readability. That's not a big deal if your reader must do it one time in your 20-page document—however, it's a giant deal if your reader has to do it for every sentence. You've added a lot of noise. Your co-workers are not going to work that hard to understand what you're trying to say.

Always write in active voice. If this is hard for you, practice starting all your sentences with "You can…" (You can print the document). It's almost impossible to write a passive voice sentence when you front-load the sentence with the actor. Later, as you get better at this, you can start sentences using the imperative (Print the document) and other methods.

The only typical exception is if you're writing customer-facing error messages for products. In this case, write in passive voice because if you put the actor at the front of the sentence, you're probably blaming the user for the problem. And while your user may have caused the problem by pressing the wrong button, nothing is helped by implying they are stupid or incompetent.

For example:

You typed the wrong password. (active voice but a little blame-y)

The wrong password was typed. (passive voice but not blaming)

You see how, although the active voice is clearer about who did what, it blames the customer for what's happened. In this case, it's better to make it less clear about who did it and focus on what happened. Even better is to provide a solution to what happened. For example:

The wrong password was typed. Check your CAPS lock.

Present tense

Tense is how we use language to express when something happened. English has 7 or 8 tenses; opinions vary, it seems (see Figure 2.1).

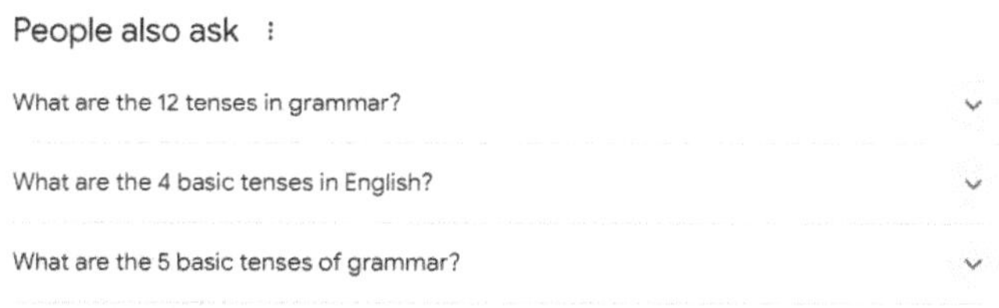

> People also ask　⋮
>
> What are the 12 tenses in grammar?　　　　　　　　　　　　　　⌄
>
> What are the 4 basic tenses in English?　　　　　　　　　　　　⌄
>
> What are the 5 basic tenses of grammar?　　　　　　　　　　　　⌄

Figure 2.1 – Common questions about tenses in English

Other languages have more tenses or fewer tenses or no tenses. It's the nature of languages.

In technical communication, we care most about the present tense. Present tense puts the action in the now, making the information immediate and relevant.

For example:

- **Present tense:** Bob runs to the car.
- **Past tense:** Bob ran to the car.
- **Future tense:** Bob will run to the car.

Notice how present tense makes you feel like Bob is running right now? You can almost picture it in your mind. What's going to happen when he gets there? Why is he running? It's inherently interesting because it's happening right now.

In the past tense, Bob may have run yesterday, last week, or several years ago. We don't know. We assume things worked out in some way because we didn't hear anything about it when it happened. It's already over. We're not that interested in it.

In the future tense, we don't know when Bob is going to make that famous run. Is it now? In a few minutes? Perhaps later? Do I have time for a nap? I have things to do. Could someone maybe call me just before Bob runs? It's not that interesting for us to think about because there is no immediacy about the event. We have time; it's later, not now.

The future tense implies that you have no idea when something is going to happen. You're adding ambiguity to the timing of the event. For example, if you're writing about a product, you might write something like: "The report will print." You're saying that the report will print eventually (who really can know when). Perhaps today, maybe tomorrow. It's just hard to know. Call us if it works for you—we're still waiting.

You don't want people to feel uncertain about your products. You want them to be certain that they know how your products work. If you don't know when your product works, who does?

For example, let's say you use future tense in a functional specification that defines how a report gets generated. You accidentally left wiggle room; your report could print overnight or next week, even if you (or, more to the point, your customers) want the report to print immediately. We do not use this example randomly, as we have seen much upset in projects over this sort of miscommunication because everyone had a different vision of "when." Using present tense and specifying the time period within which you expect the report to print avoids this miscommunication.

This is not to say you can never use past or future tense. But the bulk of what you're writing can and should be in the present tense. An easy way to find future tense in your writing is to search for the word "will." In English, the structure "will <verb>" is how we create future tense. So when you find the word will, delete it. Then look at the verb to see if you need to add an "s" to it.

Sadly, past tense isn't this easy to find in English because English beats up other languages and steals words from them. But this tip gets you started to find and fix future tense.

Second person

English has multiple pronouns: he, she, it, they, we, I, and you. In technical communication, you almost always want to speak directly to the person you are writing to, so we use *you*. It would be odd to speak directly to a friend and use another pronoun or, worse, "the user." You can imagine asking that attractive person out for coffee and referring to them as "the user." "Would the user like to go out for coffee? I bet they have the muffins the user liked last time!" That's not going to result in a nice chat, and it might get you reported to someone (or blocked) as just strange.

The same goes for your technical communication. You're writing to a specific person or people; consumers of your document or the engineers building the product. Because you're talking directly to someone, use *you* when you need to refer to them.

That said, sometimes, you do need to refer to "the user" when there is another group of people who are not the people you're writing directly to. For example, if you're writing a specification for a systems administration product, there are two layers of users—the *sys admins* (your audience) and their *users*. Some people are, in fact, users of that product—the people who need the sys admin to create accounts and set up permissions. In this case, you do refer to them as "the user" because they are *your readers'* users.

Lastly, in English, "you" is more familiar and more trusted at an unconscious level. If you're writing in other languages, the familiar "you" may not be appropriate. For example, in Spanish, the familiar you (*tú*) is generally reserved for family and close friends. Using it in technical communication where your audience is strangers is presumptuous. And other languages have different versions of *you* that make it all more complicated.

Repetition

If you call something a widget in your first paragraph, use that word for that thing all the time. In academic writing, you were taught that using the same word over and over is redundant and potentially boring. And when you're writing to demonstrate your mastery of language and the topic, this may be true. It's certainly not true in technical documents.

Repetition is clarity. Repetition reduces confusion (noise) because everyone knows exactly what you're talking about every time you talk about it. No one gets confused because here you call it a widget, and there you call it a banana, and later you call it a cat. With repetition, you are not boring your readers—consistency helps ensure you're communicating clearly.

Repetition includes repeating your noun or noun phrases instead of using "it" or "this." It's not always clear what "it" or "this" refers to, so repeating your noun or noun phrases removes that possible confusion. Repetition adds clarity to your communication.

Reduce the reading level

The average reading level in the US is 5th to 7th grade.[3] And while you may be writing for other engineers, your communication is not the only thing vying for their attention. The simpler the words you use, even when you're communicating complex technical information, the easier it is for your audience to understand and remember. Especially if, as is often the case, your fellow engineers are not native English readers. In no way are we saying you should talk down to your audience; we're saying meet your audience where they are and get your message through.

Think of it like a radio frequency (say, AM vs. FM): if the receiving radio cannot accept FM, are you "dumbing down" the signal if you adjust it to AM? Of course not. You're just changing the frequency to ensure the signal gets through.

Here are two simple ways to reduce the reading level:

- Keep words of three or more syllables to a minimum. Where possible, except for appropriate technical terms, use smaller words.
- Avoid what linguists call *latinates*. These words—like utilize (use), facilitate (help), and so on—may make you feel like your writing is more formal or more intelligent, but they can be confusing. And they add unnecessary noise to your signal.

User-focused, user-centric

Because we write to people, we need to keep the focus on the reader. That sounds obvious, but it gets forgotten. Business readers need to know why they should care about this information you're communicating.

Business readers have a lot vying for their attention, so it's up to you, the communicator, to show them why they should care. And you must show them why in the first few sentences. You cannot expect people to read your communication because they're desperate for any word from you.

Always think first about your readers when you write—what do they care about? How can you talk about your subject so that you're engaging them? The writing guidelines so far help a lot. But so too does talking about products from the reader's point of view—not the product's.

---

[3] "What is readability and why should content editors care about it?" (Marchand 2017)

For example:

Product-centric: The WidgetProduct uses tags to talk to other products.

or

User-centric: You can use tags to let the WidgetProduct talk to other products.

The first example is written from the point of view of the product. It tells us something that the product does but doesn't tell us why we should care. Oh, we think, it's nice that WidgetProduct uses tags to talk to other products. It should probably continue to do that. But, as written, it seems not to need *you* to do that. The information is presented as a factoid, but you're not told why you should care or why this information is important to you.

The second example uses the writing guidelines (active voice, second person, starting the sentence with "You can...") to connect the reader directly to what the WidgetProduct can do. The reader is involved because you start the sentence by involving them. You tell them what they can do in the product. It's not a factoid hanging in space; it's an action they can take to expand the product and what it does. There is a specific problem your reader may have, and you show how the product solves that problem or takes a step towards solving that problem.

Short is good

People in business are busy. They have meetings, deadlines, and projects to attend to. Their minds are constantly trying to process multiple things at one time. The term *cognitive load* refers to how much of their brain's capacity is occupied at any given moment. Too much cognitive load means that your reader is mentally overwhelmed with information from the world while trying to read your content.

Think of it this way—reading a complicated book in a quiet house on a relaxing weekend vs. reading the same book sitting in the middle of Grand Central Station at rush hour while you're trying to keep an eye out for when your delayed train will leave so you can go to the job interview you hope will lead to your dream job. Which situation makes it harder to understand and remember the content in the book? Grand Central Station is likely to be harder because there is more going on. All that stuff about that situation makes it harder to focus because so much of the mental capacity is already in use. That stuff going on is cognitive load.

We can't control if or when our readers are struggling under a significant cognitive load, but we can do things to avoid increasing it. In addition to following the guidelines above, the next thing we can do is keep our information short.

Short sentences

Keep your sentences to 25 words or fewer. Research shows about 20 words is a good upper limit.[4] Short sentences are easier to understand because there are fewer ideas to consume at one time. The fewer ideas you pack into a sentence, the easier it is for your reader to understand the ideas. The key thing is to vary your sentence length. You don't want all your sentences at or near the upper limit.

With fewer than 25 words per sentence, you don't have unlimited room to explain your ideas. Make every word count. Remove extra words and phrases to make room for words and phrases that are needed for your idea.

You may think this sentence-length limit is impractical, that no one can write about complicated technical issues and keep their sentences under 25 words. You'd be wrong. Take a look at any example of technical information that you can easily understand without having to re-read it several times. You will find that it probably follows this guideline.

Short sentences also prevent you from adding unnecessary words. When you have 25 or fewer words in every sentence, every word must contribute to the signal. Additionally, short sentences lend themselves to the subject + verb + predicate construction that helps people understand and remember content.

What words can you cut or change? Start with the list in Table 2.1. It's not comprehensive, but removing these phrases from your writing can go a long way to removing noise from your signal.

---

[4] "Sentence length: why 25 words is our limit" (Vincent 2014)

Table 2.1 – Phrases to cut from your writing

Not this...	Try this...
aims to <verb>	<verb>
are used to/is used to <verb>	<verb>
could <verb>	<verb>
desire	want
facilitate	help *or* support
has been	is
has been <verb>	<verb present tense>
has finished	is done *or* is complete
has the option of	can
have finished	are done
have the ability to <verb>	can <verb> *or* <verb>
have the option to	can
if you want to <verb>	to <verb>
in order to	to
is able to	can
is/are designed to <verb>	<verb>
is used to <verb>	<verb>
may	can
might	can
offers the ability to <verb>	lets you <verb>
once	one time *or* after
should <verb>	<verb>

Not this...	Try this...
since	because *or* after
utilize	use
will also be able to	can also
will be <verb>	is *or* are <verb> *or* <verb present tense>
will be able to <verb>	<verb>
will have been	is *or* are
will have the ability to <verb>	can <verb>
will have to	must
will have to be	must be
will then be	are *or* is
wish	want
without the need to access	without accessing
would <verb>	<verb>
would be	are *or* is
would like	want

Short paragraphs

As you know, paragraphs are groups of related sentences. Readers understand best with about 3 to 5 sentences in a paragraph. That means you have about 125 words before you need to change the topic slightly. If you look at your paragraphs, you probably see you are already slightly changing topics at about the fifth sentence. Break your paragraph right there. That paragraph break helps people know when the subject is going to change and gives their brains a rest.

Here's an example of a large paragraph.

> Sentences should be less than 25 words. Research shows that more than 25 exceeds the cognitive carrying capacity of working memory, and a reader's ability to remember the content falls significantly. The sweet spot seems to be 20 words. Overall, write to a 5th-grade level, even if you expect your audience to be domain experts. The average reading level in the US is about 5th grade—not because Americans are foolish or uneducated. America has a large non-native English speaker population, and this impacts the overall reading levels. Clear writing keeps sentences under 25 words and aims for a 5th-grade level. Additionally, your audience probably includes many non-native speakers. Reading in a second language while trying to follow a procedure in a complex or new tool is a lot of cognitive load. You can reduce that load by using simpler, shorter words and using the same word for the same thing. Plain English research has shown this is most effective to help people remember and act on your words.

This looks like a lot of text. Nothing about this makes you want to read it because it's a blob of text. Adding paragraph breaks makes this much easier on the eyes and easier to read.

> Sentences should be less than 25 words. Research shows that more than 25 exceeds the cognitive carrying capacity of working memory, and a reader's ability to remember the content falls significantly. The sweet spot seems to be 20 words.
>
> Overall, write to a 5th-grade level, even if you expect your audience to be domain experts. The average reading level in the US is about 5th grade—not because Americans are foolish or uneducated. America has a large non-native English speaker population, and this impacts the overall reading levels. Clear writing keeps sentences under 25 words and aims for a 5th-grade level.
>
> Additionally, your audience probably includes many non-native speakers. Reading in a second language while trying to follow a procedure in a complex or new tool is a lot of cognitive load. You can reduce that load by using simpler, shorter words and using the same word for the same thing. Plain English research has shown this is most effective to help people remember and act on your words.

Short sections

Sections are groups of related paragraphs. Research shows that readers want no more than five paragraphs before they see a section heading.[5] A section heading is a title set on its own line and distinguished visually from the main paragraphs. You can see examples throughout this book, including the section titled "Short paragraphs" and the section titled "Headings" below.

Keeping sentences, paragraphs, and sections to these limits is not just our opinion. People under cognitive load have a lot going on in their heads. When experiencing cognitive overload, the amount of information they can keep in working memory is limited. Providing information in smaller pieces makes it easier to transfer that information to longer-term memory.

Additionally, you may be sharing new information with people. People learning new information need it in smaller pieces so they can chain it all together in their minds.

Headings

Use headings to break narrative text into chunks. Headings show the logical relationship between the different groupings of content and ideas on the page. Look at the headings on this page—you see how the ideas are related by looking at the headings. Your reader can understand more about your ideas and how they are linked together by looking at your headings.

> (i) *Headings* are not *headers*. Headers are the text in the upper right and left of a printed page. Headers support navigation in a printed page as people flip through a book, looking for a specific chapter, for example. Mistaking these terms causes no end of confusion.

Headings make your content scannable. Headings make it possible for people to scan your document to find the information they need. For example, if a product manager is looking for the project's scope, she can scan for the section labeled **Scope** and ignore everything else.

Headings are also important if the document is read online. People visually scan documents, looking for what they need. For example, you wrote a troubleshooting article for the knowledge base. Headings let your readers scan the article quickly to see if it applies to their issue.

---

[5] "The Layer-Cake Pattern of Scanning Content on the Web" (Pernice 2019)

However, the research quoted above shows that putting a heading on every paragraph limits scanning and slows comprehension. It's even worse to place a heading on every sentence. It makes the page visually jarring and difficult to read. The sweet spot is between three to five paragraphs.

Headings can also make the entire page look more open and inviting. This can be critical in technical material. A more open page makes the content look easier to read and understand. Humans are suggestible, so a page that looks easy to read often turns out to be easy to read. A page without headings is typically hard to read.

> **Try this:** Open a textbook at random and place it at your feet. Do the pages look like gray blobs? Pick the book up and look more closely at the pages. Does a paragraph go on for several pages? Are there any headings after the name of the chapter? Does any of the content look like it's going to be easy to read and understand? Probably not.

People use headings to find what they're looking for. Research shows people use an F pattern to scan the page.[6] It's an algorithm: they scan for a heading that's sort of like the information they're looking for. When they find that sort of heading, they read the first sentence. If that sentence is what they're looking for, then they continue reading, else they go back to scanning.

If that's how people scan, create information that supports that method. Use headings.

Building sentences and paragraphs

At this point, you understand concrete things like active voice, present tense, how to construct sentences and paragraphs, and how to use headings. It's time to start putting it all together.

Topic sentences

Paragraphs start with a topic sentence. As you recall from high school, a topic sentence states the idea of the paragraph. You may have been told that you can place a topic sentence at the end of your paragraph. Not in technical writing.

You need the first sentence of the paragraph to tell your reader what this paragraph is about. Readers who don't know this material use that sentence to understand what's ahead in this paragraph. Scanning readers can read the first sentence and decide if this is what they are looking

---

[6] "F-Shaped Pattern of Reading on the Web" (Pernice 2017)

for or if they should move on. Either way, it's critical that your reader get this important information up front—what is this paragraph about?

Sometimes, if you're covering very technical information, it's possible that you need to use two shorter sentences to function as one longer topic sentence. Often, this looks like:

- Complicated Concept 1 is related to Complicated Concept 2, and this changes Complicated Concept 3.

Often, it's the word *and* that's the clue that you might want to think about using two sentences as one topic sentence. How is that the clue? Because you already have a lot going on in this sentence before you get to the word *and*. After you get there, you're adding more complicated information. It's better to split this into two sentences, such as this:

- Complicated Concept 1 is related to Complicated Concept 2. Complicated Concept 2 changes Complicated Concept 3.

Notice how this example repeats the end of the first sentence at the beginning of the second sentence. You'll see another example of this in a few pages.

You should never need more than two sentences as a topic sentence. If you think you need three or more, you don't understand what you're writing about, or you're not explaining it clearly. Stop, go for a walk, take a break, and come back to it. You're trying to explain all the concepts at one time. Pick one and talk about it. In the next paragraph, talk about the next one, and so on.

Let's go back to the example paragraph we broke up earlier. This time, we've rendered the topic sentences in bold type.

You can see now the topic sentences are what the paragraph is about. If you outline before you write, these topic sentences are probably the points you would use to outline that paragraph.

Sentences should be less than 25 words. Research shows that more than 25 exceeds the cognitive carrying capacity of working memory, and a reader's ability to remember the content falls significantly. The sweet spot seems to be 20 words.

Overall, write to a 5th grade level, even if you expect your audience to be domain experts. The average reading level in the US is about 5th grade—not because Americans are foolish or uneducated. America has a large non-native English speaker population, and this impacts the overall reading levels. Clear writing keeps sentences under 25 words and aims for a 5th grade level.

Additionally, your audience probably includes many non-native speakers. Reading in a second language while trying to follow a procedure in a complex or new tool is a lot of cognitive load. You can reduce that load by using simpler, shorter words and using the same word for the same thing. Plain English research has shown this is most effective to help people remember and act on your words.

The rest of the paragraph

The other sentences in the paragraph amplify and support the topic sentence. They further explain the concept or idea you presented in the topic sentence. The paragraph can end with an example that illustrates the concept. Examples help people better visualize what you're communicating.[7]

Another way to end a paragraph is with a limitation or special case the reader should understand now that they know about the general class. Not every paragraph will end with an example or special case, but using one to end a paragraph can often clarify your communication.

You can also use bulleted lists to great effect, especially if you have a list of, for example, restrictions or special cases. For example, you can end a paragraph like this: After you create a blob, you can:

- add contacts
- group emails
- update the membership lists

---

[7] For ideas about why examples resonate and how to improve your examples, see Appendix A, *Metaphors.*.

Using bullets makes the list more visible on the page, which helps people scan the page for the list. In general, use bulleted lists for two or more items.

Example paragraph

This section looks at an example paragraph that follows the writing guidelines. The example explains how to use dotfields in a configuration system. This paragraph assumes the reader already knows about using tags and properties in the system. Read it carefully. The following sections examine what this paragraph is doing and how.

> **Example paragraph**
>
> Each type of tag has a set of unique properties, and these properties are defined in dotfields. Dotfields can access, monitor, and modify tag properties. You can use dotfields to affect your animations by accessing and modifying any of the dotfields related to the tag selected. Not all dotfields work in all expressions.

Parsing the example

Let's examine how and why this information is working well. The first sentence is:

- *Each type of tag has a set of unique properties, and these properties are defined in dotfields.*

This is the topic sentence. The end of the first phrase is repeated as the beginning of the second phrase. This repetition links these phrases (and ideas) together. Remember, this example is shown here in isolation; the reader already knows about tags in general.

Next sentences:

- *Dotfields can access, monitor, and modify tag properties. You can use dotfields to affect your animations by accessing and modifying any of the dotfields related to the tag selected.*

These sentences explain what the reader can do with dotfields. These sentences are written to the reader. We start by repeating the word *dotfields* in the first sentence. Notice the second sentence starts with the "You can…" construction, pulling the reader right into why they care. They care because there is an action they can perform. The actions are not described from the product's point of view.

The last sentence:

- *Not all dotfields work in all expressions.*

This last sentence tells the reader about a limitation of the general class. The manual's authors need to tell the reader about what dotfields can do before they tell the reader about the limitations. Of course, you now need to tell the reader about the specific limitations, perhaps in a table where you list all the dotfields.

Notice that no sentence is longer than 25 words. The longest sentence is 21 words:

- *You can use dotfields to affect your animations by accessing and modifying any of the dotfields related to the tag selected.*

Look at the amount of information in this long sentence. There are no extra words, and every word counts.

The paragraph follows all the writing guidelines discussed in this chapter. It's a strong, effective paragraph that communicates in a user-centric way.

CHAPTER 3

Writing Good Procedures

Now that you have the basics of clear writing, it's time to apply them. One of the more common kinds of information you'll write in the business world is procedures. Procedures are core to most business writing and certainly core to most technical writing. If you write a functional spec or a test case, you're very certainly writing procedures. If you work in support, you spend your day writing the steps for people to follow to solve their problems.

Good procedures, procedures that people can easily follow, are like computer code or machine code. Your reader is going to follow what you've written and probably only what you've written, so you must write procedures clearly and specifically, just as you would in code for a machine. Good procedures essentially "program" your readers to successfully complete a task. If you program them correctly, they will continue to use your product.

Continuing to use your product is called *stickiness* (how frequently the customer is in the product to do the things) and leads to *recurring revenue*. Recurring revenue means customers continue to pay money to use your products. In the world today, with Software as a Service (SaaS), for example, customers must see value in your product on a regular basis to continue to pay for it.

So this is important stuff. Providing good procedures for your products is tied to income-generating activities that keep us all employed. Additionally, poor procedures can result in an increase in support costs because customers can't figure out how to complete a task or fix a problem. As we explore in more detail in Chapter 5, *The Business Context of Communication*, revenue is "good" and we want more of it. Costs are "bad" and we want as few of them as possible. Good procedures are the cheapest way to provide customer support.

Even if you're not writing procedures for your products, you are writing them for other business reasons, like the expected workflow in a Jira ticket or testing procedures. People need to follow your procedures, or they waste time asking for clarification (wasted time is an unnecessary cost and unnecessary costs are "bad"). Or they don't do the process correctly—or at all.

When you are writing procedures, it can be challenging to know what goes in each step. The information in this chapter helps you learn and understand the basics of what procedures are and how to write useful and actionable procedures.[1]

What are procedures?

Procedures are everywhere in the business world. They're the steps to do things in a repeatable and standard way. For example, in test cases, the testers need to accomplish the test task and follow the procedure exactly as written to do the test. Or, as another example, you want to submit your vacation plans to book the days off in the human resources application your company uses.

A procedure is a series of numbered steps that takes a reader from the start of a thing they want to do to the end of what they want to do. Each individual step starts with an action the reader takes and concludes with the result of the action, if needed.

Ideally, we want to keep the total number of steps to 10 or fewer where possible. Why? Because the limits of short-term, working memory in most people is five plus or minus two.[2] This mental capacity is important because you're asking people to read the steps and then keep that information in their working memory while they act on it. Keeping the steps to a smaller number reduces the cognitive effort and helps your reader be successful.

For example:

1. To log in, type your username and password.
2. Click **Log in**. Your dashboard opens.

This example is basic, with two steps to log in. The result of the actions in step 2 are shown after the action, as part of the step. The result allows the reader to check whether they're on track and can move forward.

---

[1] Seth Godin, a leading authority on effective marketing and leadership, has an excellent blog post on underlying things to consider when writing procedures and instructions."Thoughts on the manual" (Godin 2023). After reading this chapter, consider reviewing the blog post.

[2] https://en.wikipedia.org/wiki/The_Magical_Number_Seven,_Plus_or_Minus_Two

The goal of a procedure

The reason people need a procedure is to walk through the steps to accomplish one specific task, or goal. The goal of the procedure is to do a thing. Doing that thing has a start and a finish.

As an example, your reader has a goal (request time off for next month, so they can go sit on a beach). They don't know how to do the tasks required to achieve that goal in the human resources software product your company is making. Their goal is not to "use the human resources software product." Their goal is to "go sit on a beach somewhere." The task to achieve that goal is to submit their time-off request in the system and have it sent on its way to approval.

The structure of a good procedure

Good procedures, procedures that are complete and easy to use, follow a specific structure. That structure is:

- **The name of the task the procedure accomplishes:** This is usually formatted as a heading, or at least a line of bold text on its own line. People tend to scan for procedures because they want or need to accomplish a task, and they know that procedures provide the steps to accomplish that task.

  People use procedures to accomplish a task or a goal. What's the goal your user has? What's the task they are doing? The procedure needs to be named from the user's perspective, not the product perspective.

  For example, here is a title from a product perspective: **Re-initialize the firmware**. This is not a great name because the reader is probably not thinking about re-initializing the firmware. They're probably thinking about updating the system, which might, if they think about it, include re-initializing firmware, but might not. A better name for this procedure would be: **Updating the BlobBlob** or **Update the BlobBlob software** or **Get the latest updates**.

- **A brief overview of the task:** This is usually formatted as text. It states the point of this procedure. Your readers don't always know why they need to do this task, and it's your job to tell them. This can be as simple as one sentence or as complex as five sentences. Typically, you can do this in one or two sentences, though, using the writing guidelines.

  For example: Adding contacts is how people get listed in your company directory.

  As another example: Getting the latest software updates keeps the BlobBlob machine updated to ensure it runs well.

- **Anything you need to know or can do as a result of this procedure:** This information is typically text and includes other tasks that users can do only *after* they do this task. It may not appear in all procedures, but it does appear in many. This information can help your reader understand the inherent order of tasks in your product.

  For example: Adding contacts allows you to send emails and other messages to them.

  As another example: After you update the BlobBlob, you can decode your cereal prize.

- **Before you start:** This is usually formatted as text or, if you need to list items, as a bulleted list. It may not appear in all procedures, but it does appear in many. The information in this area includes the information the reader needs to know before they start this procedure. It can also include anything they must complete before they can start this procedure, or any starting states that must happen before this procedure can be done.

  For example: Adding contacts is how people who work in your company are listed in your company directory for everyone to send email.

  As another example: When you start an update, the BlobBlob machine runs a system check and updates any out-of-date software in the system. Before you start, make sure the BlobBlob is connected to a high-speed internet connection. We recommend a high-speed, low-traffic wireless connection or a T1 line.

- **Notes, cautions, or warnings you must know before you start this procedure:** This information may not appear in all procedures. This information is typically text, sometimes with a word or an icon (or both) indicating the level of importance.

  This calls out anything the reader needs to know that's more than general *before you start* information. It can also be information that's critical to the successful completion of the procedure or to the safety of the person, data, or equipment during this procedure.

  For example:

  Before you add contacts, make sure you add child companies first. When you add a contact, you must assign them to a child company already listed in the system.

  As another example:

  While the update is running, do not disconnect power to the BlobBlob. If the BlobBlob loses power during the update, the update may not complete and you may lose data.

In these examples, we tell them first what action to take or to avoid. Then we tell them why—the consequences of that action—so they know why this should or should not happen. Always tell your reader what the results of the action are because this helps your reader understand why they should or should not do the action. It can also help them to troubleshoot when something goes wrong.

In general, notes are useful information. Cautions are information that prevents the reader from damaging the machine. Warnings are information that prevents the reader from damaging themselves or other people.

- **The actual steps:** This information is formatted as a series of numbered steps, sometimes with substeps listed below the numbered steps. This is the list of actual steps you want the reader to do, in the order they need to do them.

> **Do not** bury your steps in a paragraph. The standard is to use numbered steps with each step describing one action and the result of that action. This may end up with you having more than 10 steps. Fear not! We'll talk about how to fix that shortly; keep reading.

The first sentence of the numbered step starts with an action the reader takes. The rest of that numbered step (or most often, a second sentence) tells the result of that step, if the result isn't obvious or standard for that operating system, for example. When you need your reader to locate something and then take action on that, tell them *first* where they need to be and *then* what action they need to take.

For example,

1. On your dashboard, open the **Contacts** view.
2. In the upper right, click the **Add Contacts** button. The Add Contacts screen appears.
3. And so on…

As another example,

1. From the **Home** screen on the BlobBlob machine, press the **Right** arrow 3 times. Press **Enter**. The Update screen appears.
2. Press the **Down** arrow 2 times and select the **Update** option. The Update screen appears.
3. And so on…

The structure above primarily applies to writing procedures for an end-user audience (your customers). The structure for writing procedures for functional specs and test cases is a bit different. See Chapter 16, *Writing Functional Specifications*, and Chapter 17, *Testing Your Products*, for examples of procedures in those document types.

The complete examples

Put together, our examples look like this:

Updating the BlobBlob

Getting the latest software updates keeps the BlobBlob machine updated to ensure it runs well. After you update the BlobBlob machine, you can decode your newest cereal prize. When you start an update, the BlobBlob machine runs a system check and updates any out-of-date software in the system. Before you start, make sure the BlobBlob is connected to a high-speed internet connection. We recommend a high-speed, low-traffic wireless connection or a T1 line.

> While the update is running, do not disconnect power to the BlobBlob. If the BlobBlob loses power during the update, the update may not complete and you may lose data.

(Possibly include a drawing or photograph here of the entire screen with labels. See the next section.)

1. From the **Home** screen on the BlobBlob machine, press the **Right** arrow 3 times. Press **Enter**. The Update screen appears.
2. Press the **Down** arrow 2 times and select the **Update** option. The Update screen appears.
3. And so on…

Adding Contacts

Adding contacts allows you to send emails and other communications to your contacts. Adding contacts is how people who work in your company are listed in your company directory for everyone to send email.

(i) Before you add contacts, make sure you add child companies first. When you add a contact, you must assign them to a child company already listed in the system.

1. On your dashboard, open the **Contacts** view.
2. In the upper right, click the **Add Contacts** button. The Add Contacts screen appears.
3. And so on…

Discussion of the examples

All these parts need to be in a good useful procedure, except where noted. These parts need to be **in this order** because that's the order people expect. When people use a procedure, they use it as written. If you skip these parts or put the parts in another order, people may miss the information because it doesn't occur in the order they need and expect.

For example, if you put the result of an action in the first part of a numbered step, people read the first sentence of the numbered step, see that result, and then stop reading. They don't get to the action you need them to do next.

If you put the reason to do this procedure after the steps, people are lost, not knowing why they might need to do this procedure at all. So they don't do it, or worse, do a procedure with no understanding they needed to do another procedure first. Your reader is counting on you to organize and order the information so they can be successful in what they need to do.

You noticed some text in the actual steps is bold. Typically, bold the items in the UI you want the reader to interact with. That's the industry standard, and you should use the industry standard. Bold text draws the eye (see Chapter 14, *Constructing Explanations*). Drawing the eye lets people who are just scanning the procedure immediately see what to click or where to type.

Pick one type of emphasis (italics, bold, quotation marks). If you are inconsistent with emphasis, your steps look like a ransom note. Inconsistency also makes procedures very hard to read and adds to the cognitive load.

Using graphics in procedures

Graphics in procedures supplement or replace written information. Graphics can support your more visual input mode readers (see the section titled "What are preferred input modes?" in Chapter 15).

You can divide graphics into four main categories:

1. Pictorial drawings
2. Photographic illustrations
3. Charts and graphs (discussed in Chapter 11, *Designing Effective Presentations*)
4. Screen captures

Regardless of the graphic type you use, always show the graphic with words to support it. Introduce the graphic with a sentence or two that provides context for the graphic in the introduction of the procedure. Then show the graphic. After the graphic, consider using a table with any labels you use to explain what the labels are referring to. Then you can include any procedure steps.

If you include the graphic with no supporting information, people may not understand why that graphic is there or what the purpose of that information is. It's up to you to provide that context.

Pictorial drawings

Pictorial drawings are generally a combination of geometric shapes converging to form a representation. Most often, drawings are used to show:

- **Diagrams:** for example the schematic of a printer or other physical device
- **Networks:** diagrams that illustrate how a network is set up and connected
- **Connections:** the ways items are connected, or information flows, such as a flow chart

Pictorial drawings, such as a schematic, often introduce a large section. People tend to refer to these graphics many times, sometimes even looking them up to answer a question they have, such as "How does this *thing* connect to that *thing*?"

If you use drawings, place labels in the drawings so people can follow the drawing. Often, drawings include letters or numbers as labels with a table underneath that describes what each label refers to. In a flow chart, the labels are usually embedded in the items in the flowchart.

Photographs

Photographs can show the state of a physical item, such as a test bed. People use photographs to see and understand what a real-world thing, such as a test bed, looks like. People tend to refer to these graphics many times, sometimes even looking them up to answer a question they have, such as "Where does the voltmeter attach?"

If you use photographs, place labels in the photograph so people can follow the photograph. Often, photographs include letters or numbers as labels with a table underneath that describes what each label refers to.

Labels in photographs can be challenging because photographs have a lot of visual information and many colors. It can be difficult to make the labels have sufficient contrast to visually "pop" in the visual complexity of a photograph. A best practice is to create a white box on top of the photograph and put the black or red letter in the white box to improve contrast and visibility.[3]

Screen captures

Screen captures are used in the software world to show a screen or area of a screen. Readers use screen captures to identify they are in the right place in a procedure or to easily locate a smaller item on the screen. You don't need to use a screen capture to show every item the reader needs to click or interact with. Including that many screen captures can confuse the reader because you provide too much information in an attempt to provide clarity.

Screen captures should show the area you need your reader to attend to. It's helpful to put a box, rectangle, or circle around the area you want the reader to pay attention to, such as the calendar controls area. After you decide what shape to use, always use that shape. You're setting a visual pattern that your reader will unconsciously look for in future screen captures.

If you have a screen with many controls and fields, you can show the entire screen as the result of the step that opened it. As you instruct your reader to move to different areas of the screen, you can show a screen capture of just those areas.

---

[3] If you or a teammate are really skilled in image editing, you can also create a mask of the item to focus on, then blur or fade the rest of the photo or set the rest of the photo in monochrome.

Always place the screen capture immediately after the step that made it appear on the screen. Resize the screen capture so that it visually fits with the text in the steps. A good rule of thumb is to resize the graphic so that the bulk of the regular text on the screen capture is a little smaller than the body text. This may not be possible if you capture a large screen.

Screen captures are a unique kind of graphic, in that people don't refer to them except to verify where they are in the middle of a procedure.

Defining task paths

How do we know what to include in a procedure, especially when many products are complex and can do many things? We use *task paths*.

Your customer purchased your product to accomplish a goal, such as "organize all my professional and personal contacts information in one place." Tasks are what they complete to accomplish, or at least move towards, the goal.

With the goal of the task in mind, you're ready to start thinking about the task path through the system to get that task complete. You can start here:

1. Using your company username and password, go to OurProductExample.com/yourcompany/ and log into the BlobBlob system.
2. From the menu on the left, click **Request time off**.

Let's stop here for a moment. What we've done is help the reader get logged in with their company login information. They may not have known they needed that information, so we put it first in the sentence, making it harder to miss. And we provide the URL for the system because they may not have it handy—perhaps it's the first time they've wanted to request time off.

We gave them the critical basic information they need to start down this task path.

We've bolded the items the reader needs to interact directly with. These are usually buttons, menus, items on the screen that we're drawing their attention to because we need them to act on (click, select, tap, and so on) them.

Next, we orient the reader (From the menu on the left) and then tell them what to click to go to the page(s) to request time off. It's likely this page has many buttons and reports and other things they could look at as well. But these things are not related to the task at hand—requesting time off. Telling the reader all the other things they can also do here is confusing and not related to the task at hand.

We need to stay with the task path in the procedure. Think of it as wearing blinders—you can only see what's in front of you to the end of the task road.

Grouping and common errors

Let's continue:

1. Using your company username and password, go to <u>OurProductExample.com/yourcompany/</u> and log into the BlobBlob system.
2. From the menu on the left, click **Request Time off**. (*A screen capture here is great, including the items for step 3 and 4 circled or boxed, per the graphics discussion earlier.*)
3. Near the top of the page, click the **Start date** calendar icon. Select the date you want to start your time off. This is your first full day of vacation.
4. Click the **End date** calendar icon. Select the date you want to end your time off. This is the last day you want to be away. The end date must be after the start date.

Step 3 is all the information needed to set the start date. This is a simple form, so these are trivial steps and appear close together on the screen. We can group them in one step because they're simple. We also clarify what we mean by first day of vacation to avoid confusion.

Step 4 does the same things step 3 did, but with the end date. We also help with a common error because hands can jerk as you click things and the end date can be set incorrectly. Knowing that's the case, we help the reader solve a common issue, perhaps before it even happens.

Big finish

Let's finish these steps:

1. Using your company username and password, go to <u>OurProductExample.com/yourcompany/</u> and log into the BlobBlob system.
2. From the menu on the left, click **Request Time off**. (*A screen capture here is great, including the items for step 3 and 4 circled or boxed, per the graphics discussion earlier.*)
3. Near the top of the page, click the **Start date** calendar icon. Select the date you want to start your time off. This is your first full day of vacation.

4. Click the **End date** calendar icon. Select the date you want to end your time off. This is the last day you want to be away. The end date must be after the start date.

5. Review the dates to make sure these are the dates you want off. When you're ready, click **Submit**. You see a message on the screen.

6. Read the message and click **OK**. Your vacation request is sent to your supervisor for approval. You can see approved vacation days on the Home screen when you log in.

Step 5 helps the employee know they have a chance to review to make sure they got it right before they click **Submit**. This helps people understand they can take their time and not worry about making a mistake.

Step 6 tells them what they see on the screen and most importantly, what happens next. They know the request is sent to their boss for approval. Step 6 also tells them how to see approved requests, making them confident they don't need to pester their boss about this. And then, we need a link here to a procedure that tells them how to view approved and unapproved requests.

Did you notice that we didn't put the exact text message that appears on the screen in step 6? That's because that text is going to change as the product continues to be developed. That means it can be a nightmare to keep up to date. Why create more work? Who wants to constantly check if this has changed and then update it when it does? What's the value of that?

In technical communication, we have a concept called *suitably vague*. The exact message is on the screen and the reader can read it. We refer to that message in the text, but that text can change every single day, and we never have to update anything. Suitably vague provides enough information to get the task done without being a constant source of extra, and unneeded, work.

Now that you understand the basics of writing a procedure, let's dig into the kinds of procedures.

Overview procedures and task-specific procedures

Overview procedures are a numbered list of the big-picture steps. In procedure writing, overview steps are useful in multiple use cases:

- Setting up a new user in the corporate environment
- Configuring the email servers
- Complicated processes, such as changing the oil in a vehicle or replacing the motor in an air conditioning unit

These are processes that either the audience doesn't do often, and therefore needs to be reminded of, or are complicated across multiple systems. The overview steps can be read through as reminders of all the parts, or they can set the expectations of the overall process. These sorts of procedures are best suited for domain experts or people who are very familiar with what they're doing and just need reminders.

Task-specific steps are usually a numbered list of the steps needed for the task at hand. These steps are typically specific to the system the reader is working in. They can also be smaller parts of an overview procedure. The example in the previous section is a task-specific procedure.

Task-specific procedures are suited for both domain experts and people who are less expert in the domain. In procedure writing, you typically write for both audiences.

> **Overview procedures: example**
>
> 1. In the Manager dashboard, create **Roles**.
> 2. After you create the roles, specify the **Installation Type** for your organization.
> 3. Add the **Incoming Server**.
> 4. Add the **Outgoing Server**.
> 5. Specify the other email features your organization needs.
> 6. Confirm the servers are correctly configured by sending a test email.
> 7. Add the users to the roles you created.

Notice these steps are not task-specific, in that they don't tell you how to set up the incoming or outgoing servers, for example. These steps are appropriate for a domain expert who knows how to set up email servers; they just need to be reminded of the order of tasks to do.

Task-specific procedures: example

1. In the Manager dashboard, click **Add Roles**. The Roles screen appears.
2. On the **Roles** screen, click **New**. In the **Role Name** text box, type the name of the role you want to add. Role names can be up to 32 characters and can include spaces. Role names can't include special characters.
3. In the **Description** text box, type the purpose of this role. Descriptions can be up to 132 characters and can include spaces and special characters.
4. (Optional) From the **Company** list, select the child company or companies this role belongs to. Not all organizations use child companies. You can select these later. You can also change or delete child company associations at a later time.
5. When you're done adding this role, click **Save**. You return to the Roles screen. You can repeat this process to add as many roles as you need.

Notice these steps are very specific and assume the reader has the server setup product in front of them. They are expected to work each step, in order, as they go. Each step starts with one action to take and is followed by a result to tell them what happened, based on what they just did. This helps the reader stay on track.

Notice also that we locate where the action needs to happen before we tell the reader what to do. This reduces the cognitive load because we give each piece of information to them in the order they need it. The reader doesn't need to flip the information in their head before taking the action.

Combining both types of procedures

Now that you know about writing each type of procedure, you can think about combining them if that's right for your audience. For example:

1. In the Manager dashboard, click **Add Roles**.
   a. On the **Roles** screen, click **New**. In the **Role Name** text box, type the name of the role you want to add. Role names can be up to 32 characters and can include spaces. Role names can't include special characters.
   b. In the **Description** text box, type the purpose of this role. Descriptions can be up to 132 characters and can include spaces and special characters.
   c. (Optional) From the **Company** list, select the child company or companies this role belongs to. Not all organizations use child companies. You can select these later. You can also change or delete child company associations at a later time.
   d. When you're done adding this role, click **Save**. You can repeat this process to add as many roles as you need.
2. After you create the roles, specify the **Installation Type** for your organization.
   a. From the list on the left, click **Installation type**.
   b. *More steps here…*
   c. When you're done, click **Save**.
3. Add the **Incoming Server**.
   a. From the list on the left, click **Incoming Server**.
   b. *More steps here…*
   c. When you're done, click **Save**.
4. And so on.

Combining these procedure types chunks or groups the steps into manageable units of steps. For more details on what grouping is and how it reduces cognitive lead, see Chapter 14, *Constructing Explanations*.

You helped the domain experts because they can read step 1 and then go do it, based on their more expert knowledge. You've supported the not-yet experts by telling them the big step and the specific steps they need to take to do the big step. We call this *layering* the information because you're supporting both audience groups in one set of procedures.

By creating procedures with these substeps, we're also limiting steps to under 10 numbered steps. While the overall number of steps is more than 10, people *chunk* these steps into the top level. This helps reduce cognitive load as they do the steps. Using this layering technique, you can create quite long procedures that are still cognitively manageable to keep in short-term working memory and act on.

Typically, you write both kinds of procedures in your career, depending on the audience and the information requirements of the audience. When you're not certain about who your audience is and what they know, use the combined procedure to help both audiences.

Complex procedures

Not all procedures are as simple as the ones covered so far. Some are complex and these can be harder to write.

Occasionally, you need to write a procedure for a really complex process. In our experience, this happens more often in mechanical engineering, but it's certainly not limited to that field. Keeping in mind what we know about the cognitive load of short-term working memory, these are a special case and require thought to do well.

100 steps of procedural death

No one wants to follow 100 steps, laid out and numbered from 1 to 100. 100 steps makes a procedure look "hard," and most people avoid doing "hard." There's nothing about a list of 100 steps that makes anyone want to do that procedure.

So they won't.

At best, they'll come up with their own method to accomplish that task or, at worst, they simply won't do the task at all. This can have serious business impacts because an important business process may not get done. The cognitive load for that many steps is too high.

So how do we solve this? How do we write good procedures that people can use?

10 or fewer steps

The total number of steps in any chunk needs to fit in the readers' cognitive load. Because we have little control over the cognitive load our readers come to us with, we take actions to decrease the additional load we place on them. That's all we have.

As covered in the section titled "What are procedures?" having fewer than 10 steps helps reduce that load a lot. Research shows people can keep three to five items in working memory at any one time and act on it. So let's leverage that. Let's use that in a first pass as we write procedures. Here's the example procedure we'll work with:

1. Using your company username and password, go to <u>OurProductExample.com/yourcompany/</u> and log into the BlobBlob system.
2. From the menu on the left, click **Request Time off**. (*A screen capture here is great, including the items for step 3 and 4 circled or boxed.*)
3. Near the top of the page, click the **Start date** calendar icon.
4. Select the date you want to start your time off. This is your first full day of vacation.
5. Click the **End date** calendar icon.
6. Select the date to end your time off. This is the last day you want to be away. The end date must be after the start date.
7. Review the dates to make sure these are the dates you want off.
8. When you're ready, click **Submit**. You see a message on the screen.
9. Read the message and click **OK**. You return to the BlobBlob home page.
10. From the menu on the left, click **View pending time off**. You see your pending requests. (*A screen capture here is great.*)
11. Click the request you just made.
12. Review the dates to verify they're correct. Then click **Next**.
13. On the next screen, complete the **Reason for the request**. You can type up to 300 characters.
14. Click **Next**.
15. On the next screen, from the **Approval** list, select the name of your supervisor.
16. Click **OK** and then click **Next**.
17. On the next screen, verify the information about your request is accurate.
18. If you need to make changes, click **Update**.

19. When you're satisfied the information is correct, click **Save** and then click **Submit**. Your vacation request is sent to your supervisor for approval. You can see approved vacation days on the first screen when you log in.

This procedure involves a lot of steps and makes the entire process look hard. Few people can follow these steps and get them done correctly. So let's start fixing them. We have several options to write good procedures. We can chunk these steps by:

- area of the screen,
- trivial clicks,
- general subtask,
- or all of these.

All of these methods in one place

Let's try all of these methods to chunk the information in one example.

1. Using your company username and password, go to <u>OurProduct Example.com/yourcompany/</u> and log into the BlobBlob system.

2. From the menu on the left, click **Request Time off**. (*A screen capture here is great, including the items for step 3 and 4 circled or boxed, per the graphics discussion earlier.*)

3. Select your vacation start date. Do the following:

   a. Near the top of the page, click the **Start date** calendar icon.

   b. Select the date you want to start your time off. This is your first full day of vacation.

4. Select your vacation end date. Do the following:

   a. Click the **End date** calendar icon.

   b. Select the date to end your time off. This is the last day you want to be away. The end date must be after the start date.

5. Confirm these dates are correct:

   a. Review the dates to make sure these are the dates you want off.

   b. When you're ready, click **Submit**. You see a message on the screen.

6. Read the message and click **OK**. You return to the BlobBlob home page.

This groups and chunks the procedure steps by general task, then by area of the screen, and then by trivial clicks in the same area. Additionally, we also layer the information in the steps to account for sophisticated users and more naive readers. For example, step 3:

Select your vacation start date. Do the following:

allows readers who are more sophisticated (familiar with requesting time off) to read the first sentence and then go do it. They probably don't need the substeps (indented steps). Because they're familiar with requesting time off, they just want a reminder of the steps needed.

Our more naive readers (not familiar with requesting time off) read the first sentence. We've set the expectation of what they're accomplishing in this step. They can use the substeps to do this part of the process.

If you like, you can drop *Do the following:* That's a style choice. We prefer to keep it because we think it naturally leads to the substeps. But if we were translating to multiple languages, we might recommend dropping it because words cost money to translate.

Chunking the steps into areas of the UI and into related clicks makes it easier to follow the steps. People can keep these small pieces of information in their short-term working memory while they do the clicks. The clicks are also in the same area of the screen, probably just a few pixels apart. That also helps people because the steps are happening in the same area of the screen.

How to know what approach is right

How do you choose what's best when you're looking at writing complex procedures?

Start by chunking the related clicks. See how far that gets you. You may find you're also layering the information as you go. These are both great first steps to clarify your procedures.

Look next at what you can group. Good candidates for grouping are items that are:

- a few pixels apart
- part of the same labeled area of a screen
- on the same wizard screen

For example, if your reader leaves a screen to continue, you've reached the end of the chunk. If you have five or more substeps, consider making multiple steps, based on where the reader is on that screen. Remember, short-term working memory can hold three to five items of information, so the total number of substeps in any step matters.

Identify the natural task pause points

All processes have natural pauses where the reader is done with a subtask. These subtasks are:

- when the reader is done with a screen in a wizard, if you're writing about software, or
- has removed the cover of the motor housing, if you're writing about replacing the motor in an AC unit, for example.

The pause is where the small task on the larger task path in the technical procedure is complete. It may require the reader to click **Next** to move to another screen, or put down *this* set of tools and pick up *that* set of tools.

Review the 100 steps of death in the procedure you wrote and look for these natural resting places. When you see them, add a heading that identifies the goal of the smaller set of steps.

What should that heading say? It depends. If you have no standard for this in your company, a great start is to name it the name of the subtask, for example, **Remove the AC Cover**. It's clear to the reader what subgoal this task does. Keep those words short and name the specific task that happens in this part of the larger task path.

Another great naming structure is numbering the heading as part of the larger task. A caution here, though: do not number the steps if the entire process is going to have more than 10 chunks.

What this looks like in technical writing

Now that we have the idea in our heads, here's what this looks like in a technical procedure:

Step 1: Log into the BlobBlob system

1. Using your company username and password, go to <u>OurProduct Example.com/yourcompany/</u> and log into the BlobBlob system.
2. From the menu on the left, click **Request Time off**.

Step 2: Specify your vacation dates

1. Near the top of the page, click the **Start date** calendar icon.

2. Select the date you want to start your time off. This is your first full day of vacation.

3. Click the **End date** calendar icon.

4. Select the date to end your time off. This is the last day you want to be away. The end date must be after the start date.

5. Review the dates to make sure these are the dates you want off.

6. When you're ready, click **Submit**. You see a message on the screen.

7. Read the message and click **OK**. You return to the BlobBlob home page.

Step 3: Submit your request

1. From the menu on the left, click **View pending time off**. You see your pending requests.

2. Click the request you just made.

3. Review the dates to verify they're correct. Then click **Next**.

4. On the next screen, complete the **Reason for the request**. You can type up to 300 characters, including spaces.

5. Click **Next**.

6. On the next screen, from the **Approval** list, select the name of your supervisor.

7. Click **OK** and then click **Next**.

8. On the next screen, verify the information about your request is accurate. If you need to make changes, click **Update**.

9. When you're satisfied the information is correct, click **Save** and then click **Submit**. Your vacation request is sent to your supervisor for approval. You can see approved vacation days on the first screen when you log in.

Discussion

You can see the natural task resting places, pauses, or meta-chunks in the steps above and the breaks we made with the new headings. For each set of steps, we're in a new area of the product and doing the next chunk of steps with a specific task goal in mind. That task goal is the name for that set of steps.

For example, **Step 3: Submit your request** includes the steps needed to submit the request, but only those steps.

A common objection we have heard is that customers may skip to step 3 and then try to do that set of steps. And that's true. But we've done everything possible to help them see this entire set of technical procedures needs to be completed in this order. We labeled the task pauses with step numbers and a heading.

Most people understand they can't just jump to step 3 and start there. In this case, if they try, they quickly discover they have no way to do these steps because they're not logged in, for example.

Some procedures, like replacing the AC motor, are very long and complicated and may occur over several days. Readers can potentially lose their place in the overall technical procedures. You can help them by adding a non-numbered text block after the heading that reminds them to complete the previous steps before starting this one.

For example, you can add:

Step 3: Submit your request

Now that you specified your vacation dates, you're ready to submit your request.

1. *Numbered steps start...*

The person doing the involved technical procedure can use this reminder of where they are in the overall process. Oh, yes, they can think, I haven't specified the days I want to take off yet. I should go do that.

CHAPTER 4

Writing Tools

This chapter is an overview of the tools and technologies available at the time of writing. This is a fast-moving field, and it's not possible to publish anything that's completely up-to-date. Details in this field will change by the time this chapter goes to press.

The field of tools available to help you communicate is exploding. Artificial intelligence (AI) is changing everything on a nearly weekly basis, from reviewing medical imaging to writing programming code to helping school children learn to construct better sentences. You can, and our students sometimes do, use AI tools to write entire college-level essays or even entire books. (If you're our student, stop that.)

AI isn't going anywhere.

But what does that mean for us in the world of communicating? Just because an AI can write it for you doesn't mean that the writing is clear or audience appropriate or even correct. AI also hallucinates, meaning it makes up things that are not true, including creating "sources" out of thin air.[1]

In addition to AI, other tools exist that help you be consistent in your words and phrases. These tools are called *terminology managers*. These tools solve several business problems:

- Your company has a specific style, or voice, for all customer-facing writing.
- Your writing may be translated into other languages. Using consistent words and phrases reduces the cost of translating because translators have fewer new words to translate.

Not every company uses a terminology manager, but many companies see real business problems solved when they use them. Many terminology management tools also include an AI engine that can write the content for you, using the approved words and phrases.

And, of course, your word processor includes the basics of spelling and grammar checking to help you avoid typos and nonsense phrases.

---

[1] "What are AI hallucinations?" (IBM 2023)

Spell check and grammar check

The simplest writing tools (and the oldest) are in word processors, email clients, Slack, and your integrated development interface (IDE). Word processors and other text tools check your spelling and basic grammar as you write. In most of these tools, you see a squiggle line that shows an issue. An IDE corrects your syntax as you write, even going as far as prompting you with syntax suggestions, based on what letters you're typing.

Most text tools look for words commonly misspelled and correct them as you type. You can also customize these tools further to replace specific combinations of letters or numbers with longer text. For example, Sharon has customized Google Docs to replace her full name when she types sb with a space after it. Sharon is lazy. Done thoughtfully, lazy is good.

These tools can also flag grammar issues, such as nouns not agreeing with verbs in number. As you write and then rewrite, messing up grammar is common, even for native English speakers. Both of us appreciate recommendations for commas, as the complexities of commas in English are many. Even as professional writers and native English speakers, we don't always catch every comma correctly.

And of course, you can use freeware or purchase additional tools that enhance the basics. These additional tools often work as browser extensions to check spelling and grammar in tools that don't offer spell checking or grammar checking. Some of these tools also provide an AI engine to rewrite your content.

Sharon uses a browser-based tool that also looks for long or complex sentences and offers rewritten options. She finds it useful to identify and help fix places where she let her thoughts get ahead of clear writing. The rewrite suggestions aren't always quite right, but they're a place to start. They support her writing by identifying areas she can improve.

These features save time in the writing process and make your writing clearer. Clearer writing means clearer communication. Your co-workers understand you better, and your customers understand you better. And if you're asking someone for $50 million dollars to fund your project, clear communication gets you that much closer to the funding. Use these tools as you write.

Terminology managers

Terminology managers are not a tool that you typically use in your personal life. They tend to be expensive because, as mentioned above, the problems they solve tend to be difficult organizational problems. For example:

- Your company wants everyone to follow a style guide and use product names correctly to support any trademarks the company owns.[2] They may want a supportive and casual voice in communications for customers and potential customers. This *branding* is important to a company, and companies don't want to *dilute* that branding.
- Your content is translated into other languages. Consistent words and phrases reduce the cost of translating because there are fewer different words and phrases to translate. On average, a word costs $.25 in each language for a professional translator, and the same word or phrase used again costs less than a new word or phrase.[3]

Smaller companies don't typically use terminology managers because the return on investment isn't clear for them. Large and very large companies often use these tools because continuous training for 1,000 people or 10,000 people to correctly use the current (and changing!) approved words and phrases is hard. At that scale, the computer is better at enforcing writing standards. It can watch as people write and offer real-time corrections.

If your company implements these tools, use them. When you're working on their equipment, writing for them, creating their work products, they get to tell you what kinds of words they want you to say and how to say them. You are representing that company, and they have a right to dictate approved words and phrases.

AI tools

And then we come to a new part of the technology world: Artificial Intelligence based on Large Language Models (LLM). As we write this chapter, the possibilities of AI in general seem huge. However, many start-up AI companies are failing because, while AI is interesting, it's also, at times, a solution in search of a problem (see Figure 4.1).

---

[2] "Tips On Consistency For Trademark Owners" (Gray 2021)

[3] "Translation costs" (Timofejeva 2024)

Brenda Tent retired from living at the age of old, surrounded by family and natural causes. A librarian from birth, Brenda was an avid collector of dust. She had a sweet heart and married her high school. She loved having hobbies and helping her sons to be disadvantaged youths. She had no horses but thought she did. The church gave her a choir because she sang like bird and looked like bird and Brenda was a bird. She owed us so many poems.

The funeral will be held in 1977 at heaven. In lieu of flowers, send Brenda more life.

Figure 4.1 – Early "AI" example[4]

That doesn't mean it's not a useful tool for many problems, but it's not the solution to all problems. For example, AI is not going to write the code for an enterprise-level human resources product. At least, not yet. But it's great if you need a simple JavaScript function to embed in a website form or a regular expression.

AI is a tool in your toolbox, much like a calculator.

For this book, we focus on AI that solves problems related to communicating in the workplace.

---

[4] *I Forced a Bot to Write This Book* (Patti 2020). A bot did not write this example; the author, Keaton Patti, did. However, it does illustrate the kinds of hallucinations that AI can generate. It is also notable because it became a widely shared meme by people who mistakenly thought it was an actual example of AI run amok.

Writing AI prompts

One of the hardest parts of using AI tools well is creating a prompt that gets you the result you want.[5] "Write a support email to a customer." isn't going to get you a great result. At best, you will get something generic with lots left for you to fill in, such as the following:[6]

> **Subject:** Support Assistance with [Issue/Product Name]
>
> Dear [Customer's Name],
>
> Thank you for reaching out! We're sorry to hear that you're experiencing issues with [briefly describe the issue/product]. Our team is here to help resolve this quickly.
>
> Could you please provide more details, such as any error messages or steps you were taking before encountering the problem? This will help us diagnose the issue more effectively.
>
> In the meantime, you can refer to our support page [link] for further troubleshooting.
>
> We look forward to assisting you.
>
> Best regards,
> [Your Name]
> [Your Position]
> [Company Name]
> [Company Contact Information]

Not great, but you at least have the structure for this email. It's a place to start.

Try to be more specific in what you need. For example: "Write an introduction to a customer that I'm solving a support issue for." is a better prompt. It gets you the following:

---

[5] For details on writing good, useful AI prompts, see "Getting started with prompts for text-based Generative AI tools" (Harvard University 2023).

[6] These examples were created with ChatGPT 4.0 in September 2024.

Subject: Introduction and Support Assistance for Your [Product/Service] Issue

Dear [Customer's Name],

I hope this message finds you well. My name is [Your Name], and I'll be personally handling the resolution of your support issue regarding [brief description of the issue]. I'm here to ensure we resolve this as quickly and efficiently as possible.

If you need further assistance or have additional information, feel free to reply to this email.

Best regards,
[Your Name]
[Your Position]
[Company Name]
[Contact Information]

Better, for sure, and less to rework. But you can do better than that if you're more specific.

For example: "Write an introduction to a customer that I'm solving a support issue for that involves our API not sending tokens correctly to their app." That gets you the following:

Subject: Assisting with Your API Token Issue

Dear [Customer's Name],

I hope you're doing well. My name is [Your Name], and I'll be assisting with resolving the issue where our API is not sending tokens correctly to your app.

I understand how important this functionality is to your operations, and I'm committed to finding a solution as quickly as possible. I'll keep you updated throughout the process. If you have any questions or additional details to share, feel free to reach out.

Best regards, …

This is much more specific and includes far less for you to rework. That may be everything you need to get you moving in the right direction.

The more specific you are, the better the results you're going to get. AI is very good at standard sorts of communication. It's not very good at other types of writing. The data sets the AI are trained on are limited.

> If you're still in school, always follow your school's rules about appropriately using AI in your work. You do not want ethics violations on your record.

AI use cases

AI is useful to provide you with a first draft of something you want to write, especially if it's a standard communication, as the examples in the previous section showed. The first draft idea is important because you shouldn't just copy and paste the AI-generated email to the customer without verifying it makes sense, at least. But if your writing process for everything is to blankly stare at the screen for a long time and then peck out a word, a first draft is valuable.

Recently, one of us needed quick definitions of many cybersecurity protocols and acronyms. We asked ChatGPT to generate them for us, saving hours of research and writing time. Of course, we verified that they all made sense and were accurate.

Here's another use case. In marketing, it's very common to create multiple emails, social media posts, and/or ads for a campaign. It can be time-consuming and more difficult than you might think to rework each one to sound a little different, but not *too* different. Repetition matters in marketing, so having those slightly different variations can really make or break a campaign. A valid use case for AI in this instance is to write one email, one social post, and one ad copy, then drop each into an AI prompt to "give me five different versions of [text]."

Another use case comes from a professional friend whose programmers put their code into a corporate-approved AI and ask it to create code comments. He reports the code comments are more complete than they've ever been, and people are thrilled.[7] The company is creating valuable and needed content by using AI to perform a tedious task. Before you do this, make sure your company approves, since it may violate confidentiality in your use of intellectual property.

---

[7] Eric Ray, July 2024, personal communication.

A use case that AI is brilliant at is "Teach me about…" prompts. Use these to get a quick education on a topic you need to learn about. You can use further prompts to dig deeper.

If you need to create a video about a topic, AI can create the video structure and the script for you. This can save you hours of creating the draft. Now you can focus on filling this structure out for specifically what you want to cover. Then you can use an AI tool to turn that script into a video, complete with human-looking people and voice-overs.

Some cautions

Regardless of the use case you use AI for, use AI-generated content only as a starting point. AI hallucinates. It makes up information. You must verify the accuracy of what it writes. If you don't, you run the risk of turning in, or working from, bad data. Bad data can, at best, make you look foolish. At worst, it can tank a project, lead to significant rework, or get you fired.

For example, in our class, our midterm exam asks the students to watch a specific one-hour movie and then write about the course concepts they see in the movie. In a recent quarter, several students used AI to generate their midterm exam. Which wasn't good for multiple reasons, not the least of which was this violates both our class policies and the university's policies. Worse, AI hallucinated scenes in the movie that simply don't exist. The students then discussed the class concepts shown in the hallucinated scenes.

The students obviously thought that because the script for that movie was in the LLM, the AI-generated content was accurate. They didn't bother to watch the movie and turned in a midterm exam describing detailed, complex scenes that simply don't exist in that movie. As we graded the midterm exams, at first we thought we'd had a stroke or other serious cognitive event. We read detailed discussions of scenes we had no memory of, despite knowing the movie well. We stopped grading the midterm exams and watched the movie again. **Hallucinations!**

These students used bad data to present and defend their case. They didn't watch the movie to know it was bad data. They just turned it in and thought they were fine. Their midterm exams got the grade they deserved. Bad data.

It's important to understand that both AI and humans make mistakes. But humans and AI make mistakes differently.[8] Humans have systems already in place to account for human error. Much more work is needed on identifying and overcoming AI mistakes.

---

[8] "AI Mistakes Are Very Different From Human Mistakes" (Schneier 2025)

CHAPTER 5

The Business Context of Communication

Technical communication, whether practiced by engineers or dedicated technical communicators, happens in a context, usually academic or business. You have already encountered the academic context in your career as a student. What you may not have encountered yet, unless you have taken business classes or worked at a job other than retail, is the business context. This chapter discusses the business context of engineering and technical communication.

Before we begin discussing how business operates, let's review a basic concept from Chapter 2: how you write differently for business readers than for your professors.

Writing for professors/peer reviewers

When writing for your professors or peer reviewers, your audience knows as much or more than you do about your subject. You're demonstrating your mastery of a topic or how new research in your area (often research you have performed) builds on existing knowledge to create a new understanding of the subject.

While clear writing is still important, you can make several key assumptions about this audience:

- They understand the basic principles and jargon of the topic.
- They are comfortable with longer sentences and paragraphs, although you should still work to minimize the complexity of your writing.
- They are comfortable with, and often require, more elevated vocabulary.

If you choose to go further into academia and pursue "blue sky" science,[1] you may never need to write outside this genre.

However, if you want to be famous and on TV, have lots of followers on YouTube, or function in the modern work place, now you're translating science and technology for the general public. Read on.

---

[1] "What is 'Blue Sky Science'?" (Labmate)

Writing for your boss/business peers/the public

By contrast, when writing for your boss or business peers, your audience knows as much or (hopefully) more than you do about their jobs, but they may not be engineering experts (that's your job). You're writing to inform or educate them about a specific topic, usually one about which they must make a decision—often a critically important one. All your previous experience leads you to believe that you can communicate with them in the same way you've always been taught. Unfortunately, that assumption is wrong and will get you into trouble.

Clear writing, following the guidelines discussed in Chapter 2, becomes essential, and you must change your key assumptions about your audience:

- They may not understand the basic principles or jargon. Indeed, they may be in a completely different field, such as marketing, finance, sales, or business leadership, or they may even be your end-users/customers. Explain concepts and avoid or explain jargon.
- They have neither the time nor the available cognitive load to parse longer sentences and paragraphs. Keep your sentences and paragraphs short and to the point.
- They may be comfortable with an elevated vocabulary but, as noted above, do not have the time to wade through it. Keep your words simple.
- They may be very smart people who are managing in a 2nd (or 5th!) language. Give them writing that makes it easy for them to understand.

So what is the business context?

Business writing is performed, as the name implies, in a business. Most businesses exist for one reason only: to make a profit (or at least enough money to keep running). While there are not-for-profit and mission-driven businesses, those aren't the kinds of businesses this section is about.

The businesses we discuss here make a profit (or break even) by selling a product for the highest reasonable price and controlling costs while doing so. When you're on the receiving end of cost management, this can seem unreasonable, but it is, in fact, responsible business behavior—when it's done right, anyway, but that's a topic for another day.

Within a business, most jobs, projects, initiatives, *ideas* rely on the flow of money through a business to function. To understand this flow, first you must understand what money and profit really mean.

What is money?

Before we start talking about how money flows through a business, there's a foundational economics concept to understand: *Money is value.*

It doesn't matter whether you measure value in dollars, yen, euros, goats, or blankets. However, dollars, yen, or euros are simply *currency*: the abstract "score" we use to talk about value. Goats and blankets are examples of a more concrete "score." What does this difference mean?

Before we as humans had currency (easily portable "value"), we bartered. The value → value equation was more concrete. I have a goat that I don't need, and I'm cold. You have a blanket you don't need, and you're hungry and want some goat milk (no goats were harmed in this example). So we trade. The exchange rate is clear and fairly easily agreed upon.

It's hard, though, to cart a bunch of goats or blankets around with you when you go to the market. So we invented currency to serve as an abstract marker or value. We have come to believe that currency=value, but it's not really true.

You might have a million dollars (or yen or euros) in the bank, but when the zombie apocalypse happens, chances are all that currency won't do you much good. No one wants the abstract "score" anymore. They want concrete "scores." You won't make a profit from your abstract currency.

What is profit?

One of the key aspects of making a profit is value (not currency). Simply put, any voluntary exchange where you trade something of less value to you for something of more value to you earns a profit. Back to our goat example. I have a goat that I would like to keep, so I can earn value from its milk and hair. But that's down the road. I have three other goats, and I'm cold *right now*.

You have a blanket. Actually, you have three blankets, but you only need two of them. You've been wanting a goat so you can have goat milk *whenever you want*.

When we agree to trade, I get the value of a blanket so I'm not cold (my profit). You get the value of a goat (and its milk) so you can have goat milk whenever you want (your profit).

We both profit.

This does not mean that the item you trade for is something you don't value. Maybe I really like that goat, or I have an eye toward the future. But the problem is I'm cold right now. Future goat hair or milk is not going to help me. So I profit.

Maybe you could keep trading blankets for goat milk. But that means you only get goat milk when you make a blanket, take it to the market, and trade for goat milk. But the problem is that you want goat milk when *you* want it, not when the market is open or you happen to have or make a spare blanket. So you profit.

What does this mean for you, today?

Think about this in modern terms. You value your time but don't have any currency. You need currency (the agreed-upon *score*). A company values your expertise and a work product your expertise allows you to create. The company has currency. The company offers you a trade: your time and expertise (and your work product) for their currency.

Or, looked at another way: you have currency. You value it, but you're hungry and value food *more*. The grocery store has food. They value it, but they want currency *more*. So you offer the grocery store a trade: your currency for their food.

This obviously applies to any business making goods or providing services. As a consumer, you want a thing or need a service that you value more than the currency involved. The company values your currency more than the good or service. As a worker, you value currency more than your time or expertise. The company values your time or expertise or work product more than the currency. This is the essence of profit in our modern, currency-based world.

Now that you understand that money is value, currency is how we measure value, and what profit really means, let's talk about how you get there. From here on out, I use "money" as a shorthand for "the agreed-upon exchange value of currency." Let's face it: Life is short, and who wants to type or read "the agreed-upon exchange value of currency" over and over?

We want you to note what we did here. "The agreed-upon exchange value of currency" is, in fact, the more *technically correct thing* to say. As experts in our field (whether that's engineering, or economics), we value the technically correct thing. However, the technically correct thing is usually a long or complicated phrase that adds cognitive load. So, because you are probably not an economist, we adjusted our vocabulary to make more sense to you. We did *not* "dumb this down." We simply used language more familiar to you. *That* is a key point of Chapter 2.

What profit looks like in a business

Figure 5.1 provides a rough and very high-level diagram of how money flows through a business.

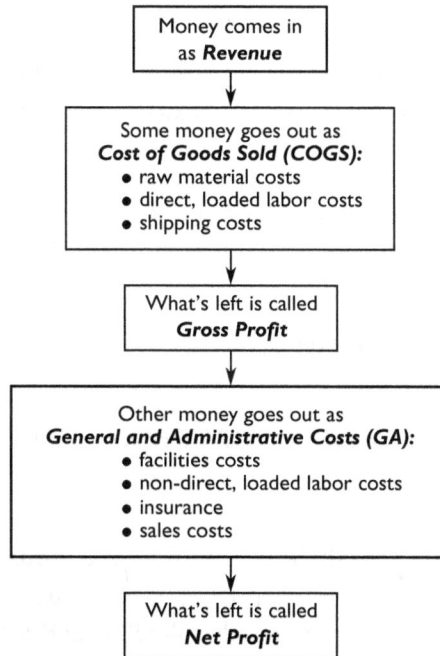

```
┌─────────────────────────┐
│   Money comes in         │
│   as Revenue             │
└─────────────────────────┘
            │
            ▼
┌─────────────────────────────────┐
│  Some money goes out as          │
│  Cost of Goods Sold (COGS):      │
│    • raw material costs          │
│    • direct, loaded labor costs  │
│    • shipping costs              │
└─────────────────────────────────┘
            │
            ▼
┌─────────────────────────┐
│  What's left is called   │
│  Gross Profit            │
└─────────────────────────┘
            │
            ▼
┌──────────────────────────────────────┐
│  Other money goes out as               │
│  General and Administrative Costs (GA):│
│    • facilities costs                  │
│    • non-direct, loaded labor costs    │
│    • insurance                         │
│    • sales costs                       │
└──────────────────────────────────────┘
            │
            ▼
┌─────────────────────────┐
│  What's left is called   │
│  Net Profit              │
└─────────────────────────┘
```

Figure 5.1 – Flow of money through a business

The better you understand this flow, the more likely it is that you will function well in business. Why? Because odds are good you'll be working in a business, and you'll probably want your work ideas and initiatives to be funded by that business. That means you'll be talking to "money people."

You get cranky when the "money people" try to get you to do something computers or devices can't actually do. You also get cranky when they can't talk about even the very basics of your field correctly. Likewise, they get cranky when you try to get money to do something it can't do. They also get cranky when you can't talk about the basics of their field correctly.

And depending on their role and place in the company hierarchy (see Chapter 9, *Flow of a Project in a Company*), they can refuse to fund your ideas. In fact, cranky people don't fund much of anything, so you don't want to make money people cranky.

Let's look at this flow in more detail. Note that we're not going to break the detailed math out for you here. There are many examples available to you with a quick internet search.

Revenue

Revenue is the easy one: it is money that comes in from the sale of your product(s) or services. There are other types of revenue, but for our purposes here, money from product/service sales is the most important.

For example, if you sell wooden birdhouses for $10 each, every $10 you get from each birdhouse is revenue. So if you sell 1,000 birdhouses, your revenue is $10,000.

$10,000.00   Revenue

Excellent! Wood-fired, artisan pizza, and craft IPA for everyone! Well…not so fast.

COGS/variable costs

Unfortunately, you don't get to keep all the revenue from your birdhouses. You need to spend some of it on what's called *Cost of Goods Sold (COGS)*. Even if you're a consultant selling services, you still have these expenses, and they're typically still called COGS. COGS includes things like:

- **Raw materials:** the wood, glue, nails, paint, and so on that you use to build your birdhouses
- **Assembly labor:** the wages you pay the people who assemble the birdhouses
- **Sales commissions:** the percentage you pay sales people based on how much they sell
- **Shipping:** the cost to send the birdhouses to your customers, either directly or through a store
- **Some taxes:** taxes that are directly related to providing the good or service, such as sales tax (consult a tax professional if you want more details)

Another name for COGS is *variable costs*. They're called *variable* because how much you spend overall changes (varies), depending on how many you make. You discover COGS fairly directly: you examine the specific costs for each of the above. For example, you bought a box of nails: how many of them did you need for one birdhouse? As another example, how long did it take to assemble? What are you paying that person per hour? And yet another example: does sales tax apply to this transaction? And so on.

Continuing our example, if you make 10 birdhouses a month, your total COGS is X, but if you make 1,000 birdhouses a month, your total COGS is more than X. Note that your total COGS for 1,000 birdhouses may not be 100X. Your materials may be cheaper per unit because you're buying in bulk. You made 1,000 birdhouses and bought a month's worth of materials at one time, so your cost per nail, glue, wood, and so on is lower. Each birdhouse cost you $2 in COGS.

$10,000.00	Revenue
$2,000.00	− COGS

Gross profit

Gross profit is what you have left over after you subtract COGS from your revenue. It's one of the measures of how operationally healthy your company is. Operational health comes first, because if you can't make your product for what you're selling it for, you will never see a net profit.

You sold 1,000 birdhouses ($10,000 revenue) in a month, and your total COGS is $2,000 (about $2/birdhouse). Your gross profit is $8,000 for that month.

$10,000.00	Revenue
$2,000.00	− COGS
$8,000.00	= Gross Profit

Gross profit *margin* is the percentage of your revenue left after you subtract COGS. For example, our gross profit margin here is 80%. For most industries, there's a range of expected gross profit margin. If your company falls within that range, you have a healthy company. If you're too far above that range, you may be charging too much for your birdhouses (customers will rapidly find a cheaper alternative), cheating, or buying substandard materials (customers will rapidly find a more reliable solution). If you're too far below that range, you may be paying too much for your materials or charging too little for your product (both of which are bad business management).

Still really good, right? So basic 3-topping pizza and Bud beer for everyone!

Overhead/GA/fixed costs

Sadly, you're not done spending revenue. You also must account for your General & Administrative (GA or overhead) costs. GA includes things like:

- **Staff salaries:** management, reception, birdhouse designers, and so on
- **Utilities:** electricity, phone, internet, and so on
- **Rent:** factory and office space
- **Some taxes:** property tax, income (profit) taxes, taxes from previous periods, and so on
- **Advertising and marketing:** publication ads, pay-per-click social, agency fees, and so on

Another name for GA is *fixed costs*. They're called *fixed* because how much you spend here stays the same (is fixed), regardless of how many birdhouses you made (or sold) this month.

> (i) Electricity, advertising, and marketing costs are not truly fixed costs, because how much you spend each month may vary slightly. For example, if you have a big demand and have to run the factory extra hours for a month, your electricity goes up. Or, if you're under capacity, you may choose to spend more on advertising and marketing for a month. However, most businesses count these items under overhead or fixed costs, because they must be paid every month.

Following our example further, if you rent a factory to build your birdhouses, you pay rent on that property whether you sold ten birdhouses, 1,000 birdhouses, or zero birdhouses this month. Same for the electric bill, the phone bill, the bill from your Internet provider, and so on.

You have a factory and an office, and keep a team of birdhouse designers and other non-factory workers on salary. You must pay your salaried workers whether you sell any birdhouses or not.

Unlike COGS, G&A is not tied as easily to the cost of production. To understand how much G&A impacts the ultimate cost of your product per unit, you must take your total G&A for a period (month, quarter, year) and divide it by the total number of units you made during that period. That is called *pro-rating* a cost.

Your GA is pro-rated as $5/birdhouse.

$10,000.00	Revenue
$2,000.00	– COGS
$8,000.00	= Gross Profit
$5,000.00	– G&A

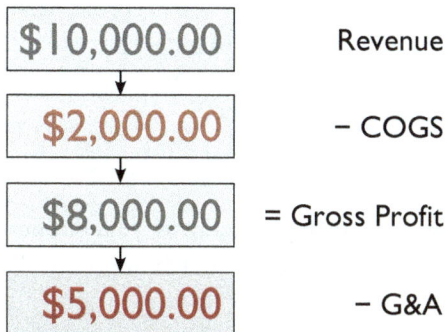

Pro-rating GA costs helps you see the true overall cost of your products and makes it easier to predict and manage net profit. If you don't do this, GA costs tend to become invisible (because you're "paying them anyway"), and that makes for real problems when you're trying to troubleshoot any financial issues in your company.

For example, many companies start cutting costs in GA first if there's a problem. But they often cut the *wrong* costs because they are just looking at the size of the overall number. We have seen companies cut all or most of their development or marketing department, for example, because the "that's a big number, let's make it smaller." Then they wonder why new products aren't getting out the door or existing products aren't updated on time or why they're not getting new sales.

You cannot "cut your way to prosperity" simply by looking at the big numbers and making them smaller. You can only do that by thoughtfully and intentionally doing the work to identify which costs are "working costs" (for example, costs that ultimately make you more profit) and which costs are just costs.

Net profit

Now we're to the part where you get to keep the revenue—well, some of it anyway. What's left after you've paid for COGS and GA is called *net profit*. It's a measure of how financially healthy your company is.

Net profit margin is the percentage of your revenue that you spend on all expenses (COGS + GA). For example, our net profit margin here is 70%. Most industries have a range of *profit margin*. If you're too far above or below that range, something bad is going on.

To finish our example, you spent $5,000 on GA costs and have $3,000 left over as net profit. That's what the business pays income taxes on, and technically, the business keeps it.

Fabulous! Firm plans for cheese pizza and generic 2-liter soda for everyone!

Retained earnings

Retained Earnings are outside the flow of money. They are like a savings account for your business: you hold a percentage of your net profit aside to pay for expenses while you wait for more revenue to come in. For example:

- The birdhouse market is competitive—you can't just keep making the same birdhouses over and over. You must innovate: design cooler birdhouses, bigger birdhouses, and so on. So you probably want to reinvest some of that $3,000 to design the next generation of birdhouses. And you're going to need more raw materials and labor to build and sell those new birdhouses.
- You have to pay your factory workers for the hours they worked, even if you didn't *sell* any birdhouses that month. You still made them, you just didn't sell them. Hourly wages are still variable costs because labor COGS varies based on how many birdhouses you *made*, not how many you *sold*. You need to use some of that $3,000 to pay them, too.

  For example, let's say you have sold all the birdhouses that are already made. Let's also say that you know you're going to sell another 1,000 birdhouses once they're made. However, they have *not yet* been sold, so you have no revenue, yet, to book the COGS against, *and* you cannot make those birdhouses without the money to pay for the associated COGS. If you don't make the birdhouses, you can't sell them, which means you'll never get the revenue to bill the COGS against. Retained earnings are how you pay for that labor *in advance* before you book the revenue. If you don't hold back some of your profits as retained earnings, you cannot build those birdhouses and book that revenue.

- You borrowed some money from a venture capital firm to start your birdhouse company. You need to use some of that $3,000 to pay them back, too.

Feel like you ended up without any money? Now you see why it's essential and ethical for businesses to keep a close eye on costs and profit margins. Otherwise, all your celebrations are "Here's your one slice of microwave pizza. Tap water's over there." Or, "You all should come watch me sip champagne and eat pizza prepared by my personal chef. Occasionally I'll waft some smell over to you while you eat the *one* snack-size Snickers bar we gave you." But don't be that person.

Communication in the context of the flow of money

You may be asking, why do I care about this? I'm an engineer. Other people handle all this money stuff. Ah, but you see, most likely you'll be making those things for a business, whether it's your employer's or your own. Either way, we presume you like to be paid for your efforts. The more you can do to make sure you engineer effectively within this context, the more likely you are to keep getting paid for what you do.

Further, as an engineer, you have ideas. You can't help it. There's little that is more frustrating than having an idea and being prevented from acting on it because you have no money to develop it. That's where understanding the flow of money can help—it teaches you how to communicate your idea so that you can get the funding to make your idea real. We'll discuss how to ask for that money in Chapter 10, *Pitching Ideas*.

CHAPTER 6
The Workplace Ecosystem

The workplace ecosystem is both very similar to and very different from the academic ecosystem. Certainly, your role in the workplace ecosystem changes dramatically from your role in the academic ecosystem.

In the academic ecosystem, you are a *student*. Your role is to learn the material your institution has determined is necessary for you to earn your degree. Your instructors are generally experts in their field, and they may be the only experts in that area of your field at the institution. You spend most of your time with instructors and other students, usually students in your major.

In the workplace ecosystem, you are a *contributor*. You are the expert in your field, at least with regard to co-workers who are not engineers. And you're typically one of many. You'll probably work with senior engineers in your specific flavor of engineering. You'll probably interact more with other engineers in other flavors of engineering than you do as a student. And, for possibly the first time, you'll interact with experts in other fields that have nothing to do with engineering but are required to get the product shipped and to keep the company running.

In other words, you're in a whole new ecosystem, with new rules and expectations for behavior.

One of the obvious things that's different is covered in Chapter 2, *Clear Writing Guidelines*. How you write and communicate with co-workers and bosses is significantly different from how you write and communicate with fellow students and instructors.

There are other things you need to know about this new ecosystem that generally no one tells you. They had to figure it out the hard way, so you have to as well. Your instructors probably didn't discuss it because they are in the academic ecosystem, and their co-worker and boss interactions still fall within that ecosystem's rules.

However, learning this new ecosystem's rules enables you to be more successful at what you came to the workplace ecosystem to do: effective (maybe even elegant) engineering while you earn a paycheck that supports the life you want.

How big is your workplace?

Your workplace may be a big company or a small company. It may support work-from-home (WFH), or require hybrid or on-premises work. It may be very structured or very loose. Regardless, most of the tips below (except where noted) apply to most workplaces.

Large vs. small companies

While the US Small Business Administration (SBA) sets different size standards for small business based on the business' classification, revenue, and a few other factors, the general definition of a small business is any "independent business having fewer than 500 employees."[1]

Small Companies

Small companies tend to be scrappy, hard-driving, flexible, and more casual. They may or may not have standardized policies, procedures, and processes for business operations. They may or may not have standardized IT rules or centralized purchasing. Typically, they don't have rigid hierarchies of titles, and in fact may not have formal titles or defined career paths at all. They have defined business needs that you can fill, but your career path is what you make it.

Don't count on a lot of work-life balance in most small companies. Because they are usually not the major player or industry leader in their market, they have something to prove. Small companies may be under-funded (especially very small companies), which means that they're chronically understaffed. The work still has to be done, so they're more likely to ask you to help with longer hours and/or more intense work.

Startup companies, by their nature, fall into this category. Startups can be very intense. The hours are long, the workload is typically impossible, and everything is fast. They are not for everyone, but there is a thrill to working in an intense and creative company that's trying to prove an idea in the marketplace. Both of us have worked in startups before and are open to it again.

Large Companies

By contrast, larger companies tend to be more hierarchical. They almost certainly have standardized policies, procedures, and processes. Those processes tend to be pretty rigid—you'd best follow them if you want a successful career at that company. They probably have a rigid hierarchy of titles. They have defined business needs *and* a distinct career path for each role that fills those needs. Generally, you have to work through each rung of the ladder to get to the next one.

---

[1] https://www.sba.gov/federal-contracting/contracting-guide/basic-requirements#meet-size-standards.

Large companies can support work-life balance better, mostly because their revenue is more regular, and there are many people to manage and perform the work. However, this is not always true; large companies can be obsessively focused on controlling costs, and in a large company, you're not really a person, you're a line item in the budget. That also means you can easily be laid off when times get tight and, unlike a small company, you probably won't see it coming.

(i) Government agencies are like large companies that have doubled down on hierarchy, defined job titles, and career paths with rigid (in fact, *legally required*) policies, procedures, and processes.

Working for a large company can be fun, especially if owning the problem space is interesting to you. It can be satisfying working for the industry leader in your area. Large companies, with established policies and procedures, are predictable, and that predictability can be comforting. Both of us have also worked in large, established companies and are open to it again.

Why do you care what size your company is?

You care how large the company is because when you are job hunting, you want to look for a company that matches your values and needs. Maybe not for your first job in your field after you graduate—you may be far more concerned with just getting a job, any job, to get the experience. And that's a valid choice. But certainly, after you have a body of experience, you probably want that match of attitude and values.

If you are freewheeling, casual, and hard driving, you may find that small companies give you more of what you need and want in a job. You may find large companies to be stifling, baffling, and aggravating. Which they can be.

If you prefer structure, support, and a predictable workday, a large company may suit you. You may find small companies to be uncomfortable, chaotic, and honestly a little crazy. Which they can be.

Neither choice is wrong or bad. It all depends on who you are and how you like to work. For example, Bonni's husband is best suited for a very small company (ideally one he owns) because he could be voted Most Likely To Start the Next Workers Revolution. He has zero tolerance for corporate shenanigans. Bonni, on the other hand, is pretty fond of that Yankee Dollar and all that that implies, so her boundaries for tolerating corporate nonsense are pretty far out there. It's just a difference of temperament.

These differences are the culture of a company. Understanding the culture of the company you're walking into helps you understand where you, and your talents, can flourish. A match of personality and culture is a joy. A mismatch of personality and culture makes for a miserable job.

On-premises vs work-from-home

Sharon and Bonni worked from home long before work-from-home (WFH) was a Thing. We both know that we are significantly more productive, effective, and frankly more fun to work with when we work from home. We have been fortunate to work with companies who understood the value of remote work long before it became more normal.

Then 2020 happened. Lockdown happened. WFH, for many companies, became the only way to keep operating. Remote tools became more plentiful and reliable. Ask us about what using Zoom, MS Teams, and so on was like pre-pandemic. It wasn't pretty, nor was it easy.

As the saying goes, it is indeed an ill wind that blows no good. Even the pandemic had its upsides, one of which was the rise of WFH. As the lockdown progressed, many companies realized they did not need bodies in the building to operate, and operate profitably.

In fact, many companies decided the *cost avoidance* of not paying for office space was such that they created a remote-first policy. Some companies are remote first, in that they don't have any office building that staff can go to, and everyone works from their homes or other locations.

What does this mean to you in your new role as a worker?

Even though, post-lockdown, many companies are trying to drag workers who don't need to be in the building back into the building, WFH isn't going away. In a WFH ecosystem, your worker role requires you to demonstrate a stronger work ethic.

In the academic ecosystem, you can probably skip class now and then, with no real penalty. For example, you won't be asked to leave the university (fired) for skipping a few classes. You can probably skip a few assignments with no more than a grade drop. In other words, the only penalty is a personal one.

But in the workplace ecosystem, especially a WFH ecosystem, "skipping class" (taking time off casually, whenever you feel like it because, after all, no one's really watching) has penalties:

- Your co-workers rely on you to do *what* you said you'd do, *when* you said you were going to do it. They may have work products that depend on your work products being complete. If you fail to deliver what you have promised or are late, you prevent your co-workers from completing their work.
- Your company relies on you to hit deadlines, many of which have a direct impact on product releases and updates—in other words, *revenue*.

The penalties for a weak work ethic are not merely personal. They now impact other people, who expect you to be engaged and productive because that's the value they're trading their currency for. And you agreed to be engaged and productive when you accepted their offer to trade your time for their currency.[2]

How to be engaged and productive

This section provides some tips that can help you both *experience* and *demonstrate* your engagement and productivity, grouped by area.

Set up a home work environment

Create a work area that's just for work. It's OK to work from the kitchen table or sofa (or bed; admit it, we all do sometimes) occasionally. But do most of your work from a defined work area. Having a work area supports better work-life balance, keeps your home and work areas tidier, and honestly, makes it easier to work and *easier to be home*. When you're done working, you walk away from that area.

We've both been privileged to have a separate room with a door we can close for our home offices. But you can create a work area even if you don't have an extra room. Try using room divider screens, or even just a special desk or small table that's only for work. When work is done for the day, put that area out of sight and ignore it.

Consider setting up that desk or table such that your back is to a wall, and you're facing out into the room. If you do that, make sure anything on the wall behind you is work-appropriate. This

---

[2] Re-read Chapter 5, *The Business Context of Communication*, if you're not sure what we mean here.

wall is also a great area to put some of your favorite art or items from your hobby to show off. Or a white board with your project notes can be helpful to you and those in meetings.

When you work from home, spend some effort reducing distractions:

- If at all possible, find a way to keep pets (think barking dogs) and children away from your work area, especially during meetings.
- Turn off your phone ringer. If you're using your phone for work purposes, leave it on at least vibrate. But many, if not most, companies that support WFH use Teams, Slack, or Zoom Workplace, which is where the majority of your work calls come from.
- Do not have the TV, your gaming system, or any other non-work devices on during work hours. The big exception is your music—but be prepared to mute that quickly if you have to take a call or join a meeting.
- Be aware of your personality quirks. If you are easily distracted, create an environment and personal rules to reduce those distractions. If even the sight of your gaming system makes you want to game, put it out of sight or work in another room. Create a rule, like, "I only game between the hours of 7pm and 10pm because that's when my work and personal tasks are complete."
- Avoid multitasking with non-work things. Yes, one of the best things about working from home is you can take a break to do laundry, make food, handle a child meltdown, and so on. The important phrase here is "take a break." Don't try to be in a meeting while you are making lunch, doing laundry, or handling other personal tasks.

Behaving professionally

In the workplace, your credibility is judged. Your boss and your peers form an impression of you, your work ethic, and your professional abilities based on both your work performance and your communication abilities.

Here are some tips for managing your professional credibility:

- Ask specific, concrete questions that indicate you are paying attention and made some effort to find out the answer yourself. Instead of asking "What did I miss in the meeting?" ask "From the transcript, I see that we decided to do X with project Y. I didn't see that there was a due date determined, and the Kanban board doesn't have that task added yet. Did we decide on a due date?" Of course, this assumes you read the transcript or watched the meeting recording or accessed the meeting minutes in some way.

- Try to find the answer yourself. For example, don't just email HR and ask what the next company holiday is. See if the company holiday schedule is available on the company intranet, on shared drives, or in a shared company calendar.

- Focus on your work product. Don't assume, but also be sensible and try to answer your own questions with the resources you have. Ask questions when necessary.

- Proofread your work—even email, chats, and work texts—as best you can given that you're likely typing fast in the moment.

- Unless you have a compelling business reason, don't argue about company processes or procedures. The company is not obligated to operate to your tastes and preferences.

Notice that we did not say "don't make mistakes." Mistakes are normal and human.

> If you tend toward anxiety, particularly if that anxiety is triggered by being "wrong," write this sentence down on a sticky note and put it somewhere you can see it:
>
> **You learn more from mistakes than from perfect performance.**
>
> Having that note visible helps prevent what we call "helpless anxiety," where you just freeze or ask amateur questions instead of taking control of your processes.

If you make a mistake, own it, fix it, and move on. Consider putting some kind of personal process in place to make sure you don't make that mistake again.

Over the course of your career, you will lose and gain professional credibility. You are (almost) completely in control of that, so take the care necessary to manage your credibility effectively.

Taking company standards and processes seriously

One of the best ways to demonstrate your ethics, be productive in the workplace, and manage your professional credibility effectively is to take workplace processes seriously, even if (maybe, especially if) you think they're dumb. This is not to say you can't suggest new, better processes; just that until those new, better processes are adopted, you need to take the existing ones seriously.

Companies (mostly) don't make the effort to define standards and processes unless they're needed. So that means that any processes or conventions present are there because those processes *solve(d) a problem*. You don't get to unilaterally decide that that problem isn't important or that the solution the company put in place is stupid and so you're going to ignore it.

File naming conventions

If your company (or instructor) has a file naming convention for work products, follow it. In our class, we don't grade assignments that aren't named correctly.

"But I worked really hard and not grading it is like I didn't do ANYTHING!" some students wail.

Our answer is always "Yes, you're right! It *is* just like you didn't do anything! If this were uploaded to a company shared drive, no one would be able to find it, which *is* just like it doesn't exist. If this were checked-in code, it would break the build because any module or function call done by someone else was expecting the correct name. You didn't use the correct name. So, in fact, it's *worse* than if you hadn't done anything because the company paid you to create a work product and no one can find it."

Abhishek Nalin · 2nd
Building Threado.com
2d · 🔇

Nobody will remember:
- Your salary
- Your fancy title
- How 'busy' you were
- How stressed you were
- How many hours you worked

People will remember:
- Your commit that caused a production issue

boredpanda.com

In engineering, working hard is not enough. Your work product must also meet company standards and the needs of the product specification you (hopefully) followed to do your work.[3]

Think about the last time you lost a file on your own hard drive. Now imagine you're working in a company of 10,000 people. In your department alone, there are over 100 people all working to get the product done. Everyone puts their documents on a shared network drive or in the Sharepoint folders. Now imagine you need to find a specific document talked about in a recent meeting that has important information that pertains to what you're coding this sprint.

---

[3] See Chapter 11, *Designing Effective Presentations*, for a fuller discussion of product specifications.

If everyone is using the default names of files, (UnnamedDocument*N*), how long do you think you'll search for that file? Is that a good use of the time your employer is paying you for? What's the Return on Investment (ROI) for that money?

Naming conventions are not about you and your preferences. They're about effectively working in your workplace ecosystem. Follow them.

Deadlines

Companies generally do not set random deadlines.[4] They may set tight deadlines, they may set unmeetable deadlines, they may set deadlines *you* don't like, but the deadlines are (usually) not random or capricious.

Remember our discussion in Chapter 5, *The Business Context of Communication*, about the flow of money? Yeah, it turns out the company can't collect any revenue for a product *they haven't released yet*.[5] Product deadlines (even interim ones) are defined so that the company knows when they can release and start promoting a product and collect revenue. Many other teams are basing their work products on your deadlines.

Meeting a deadline means that your work product is completed to specification and checked in or otherwise made available before the last stroke of the last minute of the deadline. Don't wait until the last possible second to commit your changes. Don't assume your computer's clock is set to exactly the same time as your company's. Don't assume your tech, your wifi, or the internet itself will work 100% of the time. It won't, so leave yourself some extra time.

You can certainly question unrealistic deadlines or have discussions about the scope of the deliverable. It's absolutely appropriate to negotiate any deadline, for nearly any reason. Be sure to frame your argument in terms of the business impact and business risks to the flow of money if you want to be successful.[6] What you can't do is blow off the deadline because you think it's stupid. That's massively unethical and more than a little passive-aggressive.

Meeting (or questioning) deadlines speaks directly to your ethics. An ethical engineer either meets the deadline or tries to negotiate the deadline by making a business case for changing it. They also raise issues well before the deadline, if they can see the issues coming.

---

[4] Usually. Looking at you, Projekt Red: "Cyberpunk 2077's launch, explained" (Favis 2021).

[5] Well, maybe you can, but after you don't actually release it on that date, you may have to give that revenue back, which is even worse and more painful than not collecting it.

[6] See Chapter 10, *Pitching Ideas*, for details.

Handling meetings

Meetings are a fact of the business world. Some companies are meeting heavy, others are meeting light, but you will attend meetings in the business world. You'll even attend meetings that you think are stupid and a waste of your time. But other people think the meeting is important and will notice if you're not there or aren't paying the correct amount of attention. Not attending meetings or not paying attention can be career limiting.

Use the following guidelines to avoid damage to your career as you negotiate meetings:

- Be on time, even if you know the meeting is being recorded. If you must be late, enter discretely and ask any catch-up questions after the meeting is over or by privately reaching out to a co-worker (who is not actively presenting or leading the meeting) via chat.
- Pay attention! Even if your WFH environment is challenging, you are expected to pay attention in meetings. The meeting leader and participants will not endlessly repeat themselves because you are treating the meeting like your emotional support background noise.

  If you missed something, try to get a recording, transcript, or even AI notes of the meeting. You can also ask co-workers who were there to catch up on what you missed. Be careful. Do that too often, and you get a reputation as someone who is not serious about your job.

  Asking for *clarification* is fine. Routinely asking for *repetition* is not.
- Invest in a high-quality headset with a microphone. The mics and speakers on most laptops are usually not the highest quality. Ironically, gaming headsets are typically the highest quality headsets you can find.
- Be aware of what's behind you, which is why we recommend room divider screens or having your back to a wall. Consider using a virtual background, but note that virtual backgrounds are not infallible. They're much better than they used to be, but they're not perfect.

  A former co-worker of Bonni's often worked from her vacation cabin. All well and good, but she didn't have a separate work area in the cabin. She did use a virtual background, *but*…her virtual background would fade out if there was movement behind her. This is how Bonni knows what this woman's teenage son looks like in his underwear. Bonni did not want to know this. No one wanted to know this.
- If you are on camera, at least wear a shirt that indicates that you're a professional, even if the company culture is casual. Sure, wear pajama bottoms if you like—we both frequently do that for comfort. No one can see that on camera, as long as you don't stand up. But attending a video meeting in a bathrobe, only a workout bra, or shirtless is too casual.

- Don't eat with your camera on in a meeting. WFH and constant meetings mean that sometimes you need to eat during a meeting. If that's the case, put a message in the meeting chat that you're eating and will turn your camera on when you're done.
- Mute all external devices and try to keep children and pets out of the room. Good luck with cats. Sharon's cat, Molli, often attends meetings, fluffy tail unfurled as she passes in front of the camera.
- Be careful when sharing your screen—everyone who's on the meeting can see everything that's on-screen:
  - ► Make sure any application that displays notifications on-screen is turned off or at least that the notifications are turned off.
  - ► Make sure the application and/or any document(s) you need to show are already open and waiting to be shared.
  - ► What your audience sees is smaller than what you see. Consider enlarging the display, or zooming in on your browser, if possible.
  - ► Consider sharing only the pertinent application. Most screen-sharing meeting tools allow you to select what's being shared.
  - ► If you have multiple screens, move everything you want to share to one screen. Move everything else to another screen. Be aware that one of your screens is identified as a primary by your computer and all notifications will appear on that primary screen.

A word about "Zoom fatigue"

Zoom fatigue describes the burnout about 26% of adults feel when they are camera/video-conferencing for long stretches of time.

"Video calls fall into a space somewhere in between presence and absence that we're still collectively learning about. The best ways to find balance ultimately come down to listening to your body, taking breaks when you need to, and being kind to yourself if you feel symptoms of fatigue."[7]

It is a real thing, which you know if you've ever felt the despair of having to go into *yet another* video meeting, where your actual face and movements do not change the outcome or productivity of the meeting. You likely feel as though the meeting could just as easily have not had the video element and been just as good (or bad) of a meeting.

---

[7] "Zoom Fatigue is Real" (Mindful Staff 2024)

Some companies require cameras during meetings and some don't. The article cited above has some excellent tips to prepare for and be grounded in video meetings. Read and consider those tips if you find yourself in a culture that requires cameras for all meetings.

Managing email

Over our careers, about half the meetings we've been in could have been handled with email or a phone call. That said, many companies are moving away from email to chat tools, like Slack or Teams, for conversations. However, email still plays a role in business communication.

Emails are an effective tool for detailed communication in the workplace. Emails are *asynchronous*, meaning the communication does not happen like a conversation in real time. You send an email, the recipient(s) sees it when they see it, and they respond when they respond. There may be several hours or even days of delay between when you sent the email and when the person responds.

That said, **emails are not chats or texts.** Emails are for longer communication where you provide more detailed and complete information. Emails are often (usually) not seen right away, so don't send emails that read like chats or texts. Use full sentences (and periods at the ends of those sentences!) and follow the clear writing guidelines to more fully discuss the issue at hand.

Email threads provide a trail of the conversation and are an easy way to recap/recall the conversation and check for action items due. If you attend a meeting that has lots of action items identified and decisions made, and you are the meeting lead, send a recap email, or chat message, depending on your company culture, after the meeting. This makes tracking accountability for decisions and completing action items significantly easier. Even better is to take notes in a shared document during the meeting that you share on screen. Send a link to that document after the meeting or include a link in the meeting invite.

Sending email

Make sure all the right people are on any email you send or reply to. Often this means **Reply All** is your friend—but not always. Avoid using **Reply All** for whole company emails. Think carefully before you add someone to or remove someone from an email chain.

Most email systems have an autocomplete function when you start typing a recipient name. Look at the choices carefully—many people have similar names and "productivity tools" like autocomplete are not perfect. You don't want to send an email to Sharon Bosworth, for example, when you meant to send it to Sharon Burton. Or to Bonnie Garmin instead of Bonni Gonzalez.

Use To, CC, and BCC appropriately. In most companies, these fields are used as follows:

- **To:** This directly involves or impacts you, and you probably have an action item.
- **CC:** You should know about this, but don't need to take action.
- **BCC:** You should know about this, but not publicly and perhaps not officially.

Replying to email

When you read an email, take a moment to read it again—all the words—before you respond. It's easy to miss important details when you are moving fast. The frustration of receiving a response that only answers half the questions asked or requests information provided in the original email is a real mood- and productivity-killer. If this happens all the time, it can feel passive-aggressive to your co-workers.

If an email is upsetting, wait at least an hour before responding. It's perfectly okay to write a venting response to get it out of your system, but write it somewhere separate from your email client, such as Notepad. It's too easy to accidentally send that emotional response. We have stories.

Consider setting up a personal Service-Level Agreement (SLA) about responding to chats and emails, even if only in your own mind with your own rules. For example, Bonni's SLA for business emails is "within 24 hours/one business day." Her SLA for chat is "within 5 minutes of seeing it," allowing for being away from her desk or phone. And Bonni always shares that with her co-workers. And of course, the suggestion about waiting at least an hour if the chat or email is emotionally charged still applies.

As a general rule, emojis in chat = totally fine, emojis in email = maybe don't. However, this depends on the culture. If you see many people using emojis, it's probably okay. If not, then don't do it. Model your behavior on what you observe others doing.

Chatting

Chat tools are for short, more immediate communication. They are excellent for a quick check-in and are usually used to replace phone or in-person conversations. Chats are *asynchronous*, meaning the communication does not happen like a conversation in real time. You post a chat, the recipient(s) sees it when they see it and respond when they respond.

The expectation in chat is that response time is more immediate than email. As we said in the email section, emails can take hours to days for a response, but chats generally take between under a minute to a few hours for a response.

In general, respond to chats quickly. They are intended to be a short conversation in lieu of a call. However, if the chat is emotionally loaded, or you are in a bad emotional space at that moment, hold your answer for a bit before sending. People are generally more forgiving in chat, but there are still limits to that forgiveness.

Pay close attention to which chat you post in *before* you send. If you post a comment to the wrong person, you can delete it, but not always before they've seen it. Most people mess this up some of the time (we've all done it and there's a lot of forgiveness available).

However, we have both seen and read about people who accidentally posted a complaint *about* a co-worker or manager *to* that co-worker or manager because they did not watch which chat thread they were using. There's a lot less forgiveness there.

Be aware that nothing you say in chat or email is private. Your company keeps logs. Those logs can be used against you, so behave appropriately. It's easy to become very casual when using chat tools, almost as though you were simply talking to someone in person. But in most cases, chat does not convey tone well (emojis can help, but they're not always reliable for that),[8] so be careful what you say and how you say it.

Know your co-workers and their boundaries. Chatting can, has, and does contribute to a hostile workplace environment if used inappropriately. And because chat systems and the company IT group keep and can access logs and chat history, it's no longer he-said-she-said. There's evidence.

If you would be embarrassed to see your chat become public (that is, in the news or on a social media platform), *don't say it.*

A word about group chats

One of the best things about chat tools is that you can create group chats and communicate with the entire team at one time. One of the worst things about chat tools is that you can create group chats and communicate with the entire team at one time.

---

[8] In the modern era, emojis are a language of their own, and, as languages do, they have dialect and context. Your understanding of a particular emoji may be very different than your co-worker's, which can get you into real trouble.

Here are a few tips for being professional in a group chat:

- Remember that you are in a group chat. Any posts in that chat go to the entire group, so make sure you meant to post what you said to the entire group.
- Only post things pertinent to the entire group. That can include delegating or checking up on tasks (accountability can be public). But be sure to identify the person being delegated to (usually @[name] works). There's often a lot of traffic on a group chat and things get lost.
- As with any other conversation, praise publicly, discipline/complain privately. Preferably, discipline/complain in person or on a video call, not on chat.

Communicating via ticketing systems

If your workplace uses a ticketing system (for example, Jira/Asana/and so on for project management, Zendesk or other for support and/or knowledge base tickets), you may need to use that system to communicate with your team. Typically, these systems are used *in addition to* email, chat, meetings, and so on.

Many companies that use ticketing systems to manage products, assign support tickets, or build knowledge base articles do so because they want to keep all the information in that place. Those companies often have a standard that all communication about a project or support issue be present in the ticketing system.

Keeping the project information in one location may not be easy. It's second nature to use email and chat to communicate about projects. Some ticketing systems integrate with your email and, possibly, your chat system as well. This makes it (relatively) easy to turn emails and chats into notes or even projects in the ticketing system. Others don't, and it's a big, screaming cut-and-paste extravaganza to transfer that information over.

We've seen environments with Slack connected to Jira. All project chats about a specific Jira ticket are automatically put in the Jira ticket. But that's custom work someone set up. Not all workplaces have that automation. You can't assume that your workplace will have this capability.

A way to look proactively professional is to ask about the company or team preference (the *standard*, even if it's expressed as a preference) about tracking tickets. After you know the standard, follow that preference.

Managing your work computer

Your employer provides you with a work computer to do work for the company on. This is not your computer—it belongs to the company. Do not use this computer for personal chats, social media, or other activities. Don't install and play games on it in off hours. Be careful about how you use search on that machine. It doesn't belong to you.

Under some employment contracts, anything you create on that computer belongs to your employer—even a grocery list. Upload your photos to edit in the fantastic and expensive editing software your employer gave you for work? In theory, they own the rights (and you don't anymore) to those photos the moment you upload them.

Even if your employment contract doesn't stipulate they own everything, you can lose access to that computer in a second. If you get laid off, usually the computer is remotely turned off with no warning, and you can't do anything about that. You've lost anything personal on that machine. You'll probably never get it back.

Additionally, anything on that computer is discoverable in a lawsuit because that machine is corporate property, not your property. As discussed above, that could help you or hurt you. Make sure it doesn't hurt you by keeping anything personal off of your work computer.

Leadership in the workplace

"Leadership is the ability of an individual or a group of people to influence and guide followers or members of an organization, society or team,"[9] typically toward a specific, defined goal.

Notice that this does *not* say "…but only if you have a specific title." While "leadership" *can* also mean "the leaders of a company," that's not how we're using it here. When we say leadership, we mean the act of leading, regardless of who you are or what your title is.

You do not need to have a management title to lead. Some of the best leaders we've seen did not have (and often did not want) a management title. They just knew things and shared them with people who needed to know those things. And they knew where to find any answers they did not already know.

---

[9] https://www.techtarget.com/searchcio/definition/leadership

Anyone can be a leader, even if you're only leading yourself. It is the *act* of leading that makes someone a leader.

Leaders do not sit back and wait for others on the team to do something. They figure out a plan to achieve the team's goal and work with the rest of the team to gain agreement on that plan and who will do the actions to accomplish it by when.

Leadership involves respect and communication both up and down the title hierarchy (or, as we like to call it, the "food chain"). Leadership involves both doing your own work and making sure the rest of the team has what they need to do theirs—even if you're not the titular manager.

You will encounter bad leadership (and bad management, which is not quite the same) in your career. You may work with someone with extreme trust and control issues, which usually results in micromanagement. You may work with a manager who is not a good leader. If you care about your productivity and your team's productivity, this is the time to step up or step out.

What it means to lead and/or work in a team

Good teams do not necessarily portion work out by the numbers. In a team, "fair" means equitable, not equal.[10]

Good team leaders and members:

- Identify and rely on each other's strengths
- Delegate thoughtfully and intentionally
- Communicate frequently—in fact, good teams over-communicate
- Set up—and use!—systems and structures and processes to support the team

What can you do when you have problems with a team member?

Teams can be dysfunctional, usually in some very predictable ways.[11] The first step is to always—*always*—communicate.

---

[10] "What's the Difference Between Equity and Equality?" (Annie E. Casey Foundation 2023)

[11] *The Five Dysfunctions of a Team* (Lencioni 2008)

We're particularly fond of the COIN[12] method for communicating issues. This conflict resolution method keeps the focus on facts, taking emotion and interpretation out of the conversation.

COIN stands for:

- **Context:** what is the environment of the issue at hand? What project(s) is it affecting? What outcome(s) is it preventing?
- **Observations:** what is the observed behavior? Not how you or others interpret or feel about the behavior, but the simple fact of the problematic behavior. We aren't concerned about why it's happening or how you or others feel about it—what are the observable facts?
- **Impact:** what effect is this behavior having on the team? The project? The organization? Again, focus on facts, rather than feelings. The impact can be small ("Keerti keeps forgetting to CC me on emails about our project, so I am missing important discussions about direction") or large ("Stan is DM-ing me through social media inappropriately and the discomfort is making it difficult to be productive"). Even if the main issue is that the behavior makes you feel bad or sad or angry or excluded, focus on the business outcome. For example, if you feel sad or bad or angry or excluded, that's going to have an effect on your productivity and probably also on the quality of your work. Focus on that.
- **Next steps:** what can we as a team, or we as individuals in a team, do to resolve the issue? Maybe it's "remember to always CC everyone on the team on all emails." Maybe Stan is told to leave you alone and/or maybe fired (and yes, it happens). The point is to have concrete, specific actions that team members can take *and be held accountable for.*

Always document these conversations in an email to all the people involved in the conversation. If your manager is leading the conversation, that responsibility lies with them, but if they don't do it, don't hesitate to do it yourself. If the issue is large or has implications for someone's continued employment, consider CC-ing human resources (HR) on the recap.

You don't need to wait for someone with a leadership title to have the COIN conversation. Remember what we said about leadership having nothing, really, to do with titles? You can take responsibility for your own conversations. You can be the leader needed right now.

---

[12] "The COIN Conversation Model" (Mindtools)

Goal-setting and planning

In preparing for battle I have always found that plans are useless, but planning is indispensable.
—General Dwight David Eisenhower, President of the United States (1953–1961)

Setting goals and planning actions is an essential element of working effectively and productively to achieve a result. Your plans may almost certainly blow up within 35 seconds of completing the plans, but the *act* of planning ensures that you can start and stay focused more easily.

Before you can make a plan, you must define your goal clearly, specifically, and concretely. Goals can be big, involving a whole team or company, or small, involving just you. Regardless of the size of the goal, nothing ever gets done without one.

Writing goals is easier said than done. In fact, it pays to have a system for making sure you've really thought through your goal and why you want it. We're fond of SMART goals, a goal-writing system first described by George T. Doran, the president of the Management Assistance Program.[13]

SMART stands for:

- **Specific:** what exactly do you want to accomplish? Be concrete and detailed, but also concise. If your goal is too long, complicated, or fuzzy, you won't accomplish anything. Strike a balance. For example, "I want to become a Team Lead at the company I work at."
- **Measurable:** how, exactly, will you know whether you accomplished the goal? Determine the right numerical metric or specific conditions that define what you mean by "success." For example, "I get the title change that says Team Lead, with the pay and responsibilities that a Team Lead gets."
- **Achievable:** can you (or your team, if you're setting team goals) actually accomplish the goal? It's all well and good to state "create a teleportation device," but the science doesn't exist yet, so that's not *achievable* (unless of course you're working on creating the science, in which case, maybe…). For example, "do I need additional training? What is that training? How do I get that training?"

---

[13] "There's a S.M.A.R.T. Way to Write Management's Goals and Objectives" (Doran 1981)

- **Relevant:** does the goal fit into the company desires and capabilities? Again, it's all well and good to set a goal to "create a teleportation device," but if your company is in the business of providing online testing solutions, it's not relevant. For example, "does your company have the role of Team Lead? What specific benefit is it to the company for that role to exist?"

- **Time-Bound:** by when must you accomplish the goal? Consider product release dates, quality assurance testing runways, manufacturing lead times, and so on. A time period of "I don't know, sometime, maybe" is a quick ticket to Never Accomplished Land. For example, "I want the job title of Team Lead in six months."

There are any number of excellent references out in the world with more examples and templates and advice. Search for "SMART goals" to find these resources.

After you know what your goal is, you can plan. Planning involves breaking down the overall goal into SMART steps you can take to do what you said you wanted to do. The only real difference between a SMART goal and a set of SMART steps is adding *who*, which you can easily fold into the "achievable" segment.

Exploring the specifics of effective planning is beyond the scope of this book. Again, there are any number of excellent resources out on the Internet. Search them up.

CHAPTER 7

Résumés and Cover Letters

In a perfect world, you would never have to apply for a job. Everyone would just know you and how awesome and competent you are, and job offers would constantly show up in your inbox. You'd be stopped on the street by employers wanting to add your skills to their companies. Sadly, few live in that world (although you *might*, someday down the road; it happens. Just don't hold your breath).

In the real world, companies who wish to hire you must have some mechanism to learn about and judge both your skills and your practical demonstration of them. That mechanism is your résumé and cover letter, which make a business case[1] to a prospective employer about why they should hire you, instead of someone else.

A lot has been said about cover letters and résumés in many other places. Bookstore and library shelves are filled with guides on creating the perfect cover letter and résumé. Every job hunting site has helpful articles. Almost every college has an employment office that helps current, graduating, and recently-graduated students with their job search. Most states have an Employment Development Department that helps out-of-work people starting or changing careers find work.

And you'll hear something different about the perfect résumé from each and every one of them.

The sad truth is there is no silver bullet, no magic feather, no perfect résumé format that guarantees that your résumé gets particular attention among the hundreds that show up for any given position. However, there are things you can do to increase your chances of capturing the résumé screener's attention so you can move on to the next level.

Rather than giving you the perfect format, or the right layout, or the magic content that makes your résumé a guaranteed winner, this chapter discusses some of the underlying theories behind what we've found to be effective. Of course, we also provide some practical tips.

(i) If the job listing itself contradicts anything we say in this chapter, *go by the job listing requirements*. Nothing turns off a prospective employer faster than seeing concretely and immediately that you cannot follow directions.

---

[1] See Chapter 5, *The Business Context of Communication*, for details on business cases.

Cover letters

Cover letters tell your story in a way that doesn't have to conform to the stricter résumé format. You can display a certain amount of personality, and you can directly connect the dots of the data in your résumé. But as with all business writing, the key is to keep it short and readable. One way to do that is to follow the writing guidelines from Chapter 2, *Clear Writing Guidelines*, by using active voice, short sentences, and short paragraphs. Another way is to follow the basic structure of a business letter.

Structure of a business letter

Business letters have an expected structure. As with other types of writing, there are conventions your readers expect you to follow. If you don't follow these conventions, you look amateur. Let's take a look at the elements of a business letter.

The addresses

First off, a business letter needs to include your contact information. Including this ensures that, should you impress the reader, they have a way to contact you for an interview or conversation.

The letter next includes the recipient's name and address. While, in theory, the recipient should already know their own name and address, including this information confirms that they received a letter intended for them.

Some tips on the address section of your cover letter:

- Make sure you include both your email and phone number. Different companies prefer to reach out in different ways, so make sure you provide multiple ways to contact you. Don't go overboard, though. Phone and email should be sufficient; don't include things like your Instagram username. However, your LinkedIn profile address is a thoughtful element. Make sure your contact information is on all pages of your résumé.
- Speaking of social media, now might be a good time to clean up your social media accounts. Consider removing anything that could cast you in a bad light. We're sure that the boat party in Boca during spring break your junior year was indeed epic. But perhaps pictures of you doing tequila shooters in a bikini do not convey the level of professionalism you would like it to, unless the job you want features bikinis, tequila, or shooters, in which case, you do you.

- Consider creating an email address just for your job hunt, particularly if your personal email address reflects a bit too much of your personality. ILoveBeerBongs420@gmail.com may work great for communicating with your friends, but it is probably not the impression you want to give to a prospective employer.
- Take the time to discover the actual name of a person at the company to which you're applying. "Hiring Manager" is easy, but using it shows that you could not be bothered to do even a minimum amount of research on a company. If the job listing itself does not provide a name, try the company website. If the website does not show names, search LinkedIn. Do your due diligence—you can be sure that they're doing it on you.

The date and salutation

Including a date in your cover letter shows that you created the letter specifically for *this* opportunity. Everyone likes to feel special—the organization where you hope to spend at least 40 hours of your life a week is no different.

Use the name your research uncovered in the salutation ("Dear So-and-so"). It adds a personal touch and, again, shows that you did your homework. Using a name indicates that you want to work here, not just that, dear God, you need a job before your parents kick you out of the house or you must vacate student housing.

> ⓘ If the person identifies as a woman, use Ms, not Mrs. Her marital status is not relevant. If you can't tell the recipient's presented gender from your research, use their full name with no title (e.g., "Dear Avery Smith").

The body

The body of the letter is where you tell your story. We'll cover that in more detail shortly.

The complimentary close

After you've told your story, close with a line or two about how you're looking forward to speaking with them further. Consider reiterating your contact information to provide a direct call to action.

Finally, you can never go wrong with "Sincerely" at the very end. It's a classic for a reason.

Do not include every time and date you're available for an interview. If they're interested in talking to you, they'll reach out to arrange a date and time that works for both of you.

Telling your cover letter story

The body of your cover letter is where you tell your story and provide context to the data in your résumé. Disclose as much as you like, but be careful to stay on the right side of the TMI line. Keep it professional and to the point, but let your passion for the potential job shine through.

No one is ever a perfect match for any given job listing. But the cover letter lets you explain how you can use the experience shown in your résumé to accomplish your potential employer's business goals. Be sure to tell only the relevant parts of your story, and make sure that you clearly and directly discuss how what you're saying is relevant.

Describe what your previous job experience has taught you, either about the job itself or about working in the business world in general. The cover letter is where you can explain how some of your job history may not look relevant, but in fact it is, and here's why and how.

For example, Bonni applied for a job as a social media manager at a company that focused on space travel. While she has no direct experience in such a company, she reads and follows space science and news avidly. Did she mention that in the cover letter? You bet she did.[2]

Résumés

The job search environment

The job search environment is competitive. Even in a favorable market (for example, unemployment is low or you have a specialty that's in demand), you face stiff competition. Each employer receives, on average, 250 résumés for any given job opening. Employers interview 4 to 6 applicants per job. And, of course, only one person actually gets the job.[3] While these are not terrible odds (they're way better than, say, your odds of winning the lottery or being eaten by a shark), you still need to present yourself in the best possible light if you want to be interviewed.

Professional vs intern/first job résumés

The résumé you create to get the second job in your career is different from the one you create for your first career job or an internship. For a first job or internship, the company knows they are hiring for raw, beginning talent, not for talent+experience. For these early job types, what

---

[2] She didn't get the job, although she found out later that it was because they already had an internal candidate, but had to post the job publicly per their hiring policy. It happens.

[3] "2022 HR Statistics" (Turczynski 2022)

you did in college or career training still matters. For any jobs you seek after that, your actual experience in a job matters much more.

Because you, as a student or perhaps early in your career, have access to many resources on campus to help with your first résumé, we focus here on tips that help you get the next job, after you graduate college. If you feel you're not getting the response to your résumé that you should, even if you have been in the field for a while, this discussion may also be helpful.

This structure is the one you use after your first job until you are much later in your career, so understanding it and using it helps you for some time.

Structure and content

As with any genre of writing, there are expected elements. Below, we discuss the generally expected elements of a chronologically organized résumé. There are other résumé styles, but they typically come into play much later in your career; we do not discuss them in this text.

As a general statement, for the first five years of your professional work, your résumé should be one side of one page. That limits you on how much information you can include. This is a good thing because your potential employer doesn't have the time or the interest to read many pages about all you've done in your career. You must focus the information on your résumé. The writing guidelines help you be tighter in your writing.

As you get past that five-year mark, you can expand your résumé to two pages, but never more than two pages. The information must fit on one sheet of paper. Even when you get to a late career résumé, it must fit on one sheet of paper. This is an excellent opportunity to include only the *most* relevant experience in the résumé you send for each job.

Make sure your contact information is in the header of your résumé so it appears on all pages of your résumé. If your cover letter and résumé get printed, these pages can get separated. You want the company to know what pages are yours.

Objective

The objective waxes and wanes in popularity. Like any other trend, in some years you can't do without it, while in others you can omit it safely. Of course, if the job listing specifically requires an objective (and even when it's out of favor, some do), you must include it, regardless of your personal feelings about its importance.

A well-written objective shows your focus on the company and its goals. We have to be honest: no company is particularly interested in your needs. Don't focus the objective on what you want *from* the job, focus it instead on what you can give *to* the job. Remember what we discussed in Chapter 2, *Clear Writing Guidelines*, and focus on the *value* you offer to your audience.

For example, we recommend a formula like the following:

> I am interested in *[insert the name of the position you're applying for]* at *[insert name of company]*. My *[insert one special talent or skill or interest in engineering]* enables me *[insert how your skill helps the company meet some **business** goal]*. This helps *[insert company name]* achieve *[insert big business goal—think why a company exists]*.

This is a guideline for what to include. It is not a prescriptive cure-all for the Down-Home, Gotta Write an Objective Blues. Use your judgment on whether you can combine sentences, or even write the objective differently (unless you're in our class—then we want the above with no improvisation).

Let's break down the variables in this formula:

- *[insert the name of the position you're applying for]* Be specific. The job is for a particular position, and the company went to the trouble of saying so in the job listing. Don't just say "I'm interested in a position at your company." That implies that any old job will do—maybe they want you to make sandwiches. Making it clear that you want *this* job at *this* company gives you an edge, particularly combined with the other elements of the objective formula.
- *[insert name of company]* Include the company name. Does this mean you have to customize every résumé? Yes, yes it does. Computers are a thing, ones and zeros are cheap, and your job right now is to look for a job, so take the time to do it right. It's respectful.
- *[insert your special talent or skill or interest in engineering]* Look at your special skills list (more on that shortly). Pick *one* special skill—ideally the one that is most relevant to the job to which you're applying. In around 5 to 7 words, summarize that skill.
- *[insert how your skill helps the company meet some **business** goal]* This is where you show your prospective employer that you *get* them. Though the specifics vary, business goals tend to focus on either gaining revenue or reducing costs. Think about what you learned in Chapter 5. How does your special skill, talent, or interest help contribute to the company's success by increasing revenue or decreasing costs? In around 5 to 7 words, say that.

- *[insert company name]* The second mention of the company name refocuses the objective from your skills to the company's needs.
- *[insert big business goal—think why a company exists]* Now it's time to think more broadly about why this company, in particular, exists and to show that you understand and share that goal. Most companies post their vision or mission statement on their website, usually cleverly titled something like "Our Vision" or "Our Mission." Find that statement and crib directly from it. Literally *use their words*—you know they like those words because they wrote them for the world to see.

As a real-world example, so you can see how this all comes together, here's the objective one of us would write if she were applying for the next level of the job she has now:

> I am interested in the position of VP of Marketing at Bob Corporation. My ability to articulate product value by developing high-quality content enables me to increase market awareness of Bob solutions by an average of 20%. This helps Bob ignite achievement and growth worldwide.

Special skills

Your unique skills are the secret sauce that makes you, well, you. These are the things you're especially good at. You may also be good at other things, but these are the things where you truly excel and which delight you.

Make sure the skills you list are relevant to the job, though. Bonni makes an absolutely killer Parma Rosa pasta sauce. She loves making it and has manufactured excuses to invite people over because she wants to feed it to them. It's "to die for" good, it really is.

But that fact is entirely irrelevant to her life as a technical marketer. Her company has never asked her to make Parma Rosa sauce during working hours as part of her job duties. Their loss (seriously, it's good), but she wouldn't list her Parma Rosa saucery skills on her technical marketing résumé.

A word about what are *not* special skills:

- Your ability to show up on time. That's not special; it's expected, like bathing, which you should also do regularly, but don't include it as a special skill (we've seen it in class résumés).

- Your basic competence using standard business tools. Unless you're a master Microsoft Word, Microsoft Excel, or Microsoft PowerPoint power user, don't list these as special skills. Instead, consider adding them quietly in work history or adding a section for Technology Tools.
- Your ability to work on a team. Again, like bathing, that's just expected. You *will* be working on or with a team, period. As an adult, you're expected to know how.

Work history

The work history section is where you prove, by making a business case, that you've actually done what you say you can do. This section is listed from most recent to oldest, so what you've done most recently is at the top of this section. Because your work history really did happen in the past, you can use the past tense to discuss it.

There are, as with the other résumé sections, expected elements, such as:

- The name of the company where you worked (or still work if you're still employed there)
- The month and year dates you worked there
- Your specific job title
- Your specific job duties

When describing your job duties, the key is to be specific, concrete, and measurable because you're making a business case that you're the person to hire. Let's break down those terms:

- **Specific:** Don't assume that the exact job duties for a job title are the same from company to company. They are not. List 3 to 5 essential duties in 3 to 5 bullets, using a sentence for each.

  Note that the grammar for this list is a little different from standard sentence grammar: *actor-action-acted upon*. You can omit the *actor*—grammatically, the actor is presumed to be you—and focus on the *action* and *acted upon*.

  For example, from Bonni's résumé: "Enhanced website content to improve lead conversion, resulting in a 24% increase in blog readership and a 57% increase in average time on page."

  Because this is from Bonni's résumé, she doesn't need to add "I" or "Bonni" to specify the actor. It's understood that the actor is Bonni.

- **Concrete:** Choose solid and strong active verbs and nouns, and include relevant details. Don't just say "increased trade show leads." Say what *specific actions you took* that increased them.

  Again, from Bonni's marketing résumé: "Increased the number of qualified leads from trade shows by 40% via booth redesign."

- **Measurable:** The previous items are standard and expected. What can set you apart from your competition is showing that you understand how these duties improved the company for which you did them. You do this by including a percentage metric that demonstrates your success. You can include metrics that support your success for each job duty, or as a separate item that applies to the whole job.

These metrics are typically best shown as a percentage because specific numbers lack context. For example:

- ► Increased qualified leads by 50, month over month, *vs.*
- ► Increased qualified leads by 30%, month over month

People reading your résumé don't know how many leads are usually added, so they don't know how much of a boost 50 more leads might make to the company. They lack the *business context* for that increase. If other people on the team were increasing leads by 1,000 every month, adding another 50 isn't impressive. But with the percentage, potential employers see the relative number and have context for the number of leads added.

Not all metrics need to be a percentage. But where the context is difficult to understand or might require understanding more about the business, showing a percentage is better.

Numbers (use *numerals* not words!) are eye-catching, especially when they appear in a block of text. Many engineers may have your engineering skills. All else being equal, showing that you also understand business needs sets you apart from the rest. That gets you the interview.

Some more examples:

- ► Led marketing teams to implement strategic campaigns that drove more than $50M in revenue.
- ► Developed and implemented social media strategy, resulting in a 30% increase in followers year over year.

The first bullet in these additional examples shows that sometimes a straight number is impressive enough or useful enough on its own. $50M is a pretty impressive number, even if you're applying to a company as huge as Microsoft or Amazon. In this example, using a percentage hides the achievement, rather than giving it context.

For example, saying "…drove 95% of company revenue…" is indeed a percent. But because we don't know the basis for the percent, it hides the achievement. Maybe 95% of company revenue is $100K, which is a much less impressive number than $50M. Context really is everything in a résumé and you need to provide that context to your reader.

Awards, presentations, publications, and so on

If, and only if, you have any professional awards, presentations, or published work, include them in this section in reverse chronological order (most recent first). In the heading, include only the words for the items you include. For example, if you have relevant publications, label the section Publications and list the publications.

No one cares what format you use to list the publication (MLA, IEEE, and so on), but consistently format the items. If you have many publications, awards, presentations, and so on, you may not want to list all of them so that you can keep your résumé to a consumable size. This is an opportunity to choose the most relevant and/or impressive examples.

Obviously, if you don't have any of these, don't include this section. Really! Don't include this section and then put *None* under the sections heading. A lack of awards or publications weakens your business case—don't call attention to it.

Education

We hate to be the ones to break this to you, but after your first job, most employers don't care about the details of your education. They merely want you to have had one. As all-important as your GPA is to you (and your parents) now, after you move on from your first job, what you actually did, not what you studied, is more important. If you've graduated in the last three years and your GPA was above 3.7, it may be okay to include it, but no one will miss it if you don't.

Keep the education section simple and direct. Note only the institution name, degree or certificate, and year in reverse chronological order. Don't include the years it took to get the degree (2025–2029, for example) because no one cares. They care you got the degree, not how long it took you. When you reach mid-career, remove the year. Ageism is real—although illegal in many places. Sadly, being illegal does not stop it from happening.

Ways to find a job

- Job boards, such as Indeed, Glassdoor, and LinkedIn
- Company websites
- Staffing agencies and job shops
- Personal contacts and networking
- Hiring or job fairs

An often overlooked way to find a job is staffing agencies and job shops. Either of these can be a great way to get experience in multiple fields and exposure to multiple ways of working. This section explains how they work.

Staffing agencies

Staffing agencies contract with companies to find and hire people, typically for short-term assignments. This helps companies because they don't need to spend the time recruiting and interviewing people for positions they need to fill. Frequently, these roles are filling an urgent need the company has for a short-term project. For example, a system admin person goes out on parental leave to care for a new baby. But the systems admin work needs to be done, so a contractor is brought in from a staffing agency to fill that role until the sysadmin person returns to work.

Or maybe the company has a new project, and they are not ready to hire staff with all the overhead that brings. They just want programmers to get this three-month or nine-month project done and shipped, and then the contract programmers can go away, and regular staff can take over.

As a contractor, you're an employee of the staffing agency and are paid an hourly rate. The company pays the staffing agency a different rate than you are paid. The staffing agency makes money by paying you less than the company pays them. For example, the company may pay the staffing agency $75 an hour for you to work. The staffing agency pays you $50 for every hour you work. Some staffing agencies also offer benefits, such as health insurance. You typically don't get paid vacation days or paid holidays.

You never pay the staffing agency directly to find you work. They make their money placing you in a short-term, or contract, role.

Some contract work is *contract-to-hire*, meaning there is a possibility that if the company likes you, and you do good work, they may bring you in at the end of the contract as an employee. This reduces the risk for the company because if they don't like you or budgets change by the end of the contract, you just go away. Not all contract work is contract-to-hire, but it's always good to ask the staffing agency if that's a possibility.

Job shops

Job shops are similar to staffing agencies, but more often they work with a few specific companies for specific roles. Job shops do all the recruiting and screening, and then send you to interviews with the company for a salaried job.

If you pass the interviews and the company makes you an offer, you become an employee of the company (not the job shop). The job shop is paid (usually a lump sum) after you work for the company for a specified time period, such as 30 or 90 days.

You never pay the job shop directly to find you work. They make their money from the company that hires you for a salaried job.

Personal contacts and networking

Getting a job through personal contacts, networking, or both usually makes the whole process easier. This is as close as you can get to "Everyone just knows you and how awesome and competent you are, and job offers constantly show up in your inbox." But it requires some work and a certain amount of "time in field" for you to build those relationships.

To build solid relationships for personal contacts and networking, focus conversations on helping the other person. For example, don't just walk up to someone you've just met in a networking event and ask for a job. Instead, look for some way to help them with a goal of theirs. To focus on helping, you must ask people questions about themselves (everyone's favorite topic). Then, actually listen to the answers and think about what you can do to help. Perhaps there's an article you can send them. Perhaps there's another contact of yours you can introduce to them. People remember those who help them, and they want to help back.

Job or hiring fairs

Job or hiring fairs can be a great place to find a job. You're in a room full of people who you *know* are hiring. The downside? Everyone else in the room is trying to find a job as well. It pays to do some research before you attend such fairs to find out which companies are participating. After you know that, you can do some research, identify which companies are a high priority for you, and bring customized résumés for those jobs. You can certainly bring a generic résumé to the fair in case you talk to a company you didn't think about in your research. But reserve that for handing to the companies you are less excited about.

CHAPTER 8

Ethics in Engineering

You cannot get through a single day without having an impact on the world around you. What you do makes a difference, and you have to decide what kind of difference you want to make.
—Jane Goodall, primatologist and anthropologist, National Geographic, 1974

In your engineering program, you had one of two kinds of ethics training happen to you, depending on your program:

1. A professor may have shown you the ethics statement from your specific professional association of engineering, available on their website, perhaps pointing to a few bullets, and then asked if you had questions. "OK," they said. And then they got on with a lecture on another topic entirely. They checked the Ethics box on the syllabus and moved on.

2. Your program may have required you to take an Ethics in Engineering class, where you read lots of case studies about situations in engineering, and you had to write what you would do in those cases. Then you were told what the "right" thing was, got a grade, and moved on with your program.

Neither of these approaches address what this chapter covers. This chapter takes a bigger view of ethics in the world and then in engineering, focused on your life story and how that story impacts and defines your ethics. Your life story and your path to your ethics are closely interwoven.

In this chapter, Sharon shares some personal information and uses herself as an example at times. None of this is meant to create the idea that Sharon is significantly more wonderful than you are or is a stellar example of ethical excellence. But this chapter needs examples, and she's sharing examples from her life.

Your life story: beginnings

In your family of origin, your parents created and raised you. During that process, they taught you many things, including your values and your beliefs. They installed the OS, if you will.

When you're around your family of origin, you probably all share the same or very similar sense of humor, have similar thoughts about what should happen around important family holidays, and are generally homogeneous.

That's not to say you're not an individual with your own likes and dislikes. But you're also not shocked when Mom makes the standard holiday sweet cakes, or the family gets together at a specific family member's house for certain events. You probably know your uncle's views on certain topics. You know these people, they know you, and you are all generally in agreement in how things sort of go.

Then you head off to college.

In college, you start meeting people. These people have different ideas than you do. Their mom serves different sweet cakes for holidays you may not have even heard of, much less celebrate.

You take classes and are exposed to ideas you never even considered. You learn about completely new things. You discover many ways to think about things, in ways that your family of origin doesn't know about, or at least, you don't know that they know about. You're exposed to ideas and people in ways that are new and eye-opening. It's the most intense and creative period of your life. You will be in other intense periods of change in your life, but this is the most intense period of your life.

It's changing you and your OS.

First-generation college students

If you're a first-generation college student, this can also be a time when you feel like an impostor. As though everyone else knows how to do college, except *you*. You may struggle to navigate a complex institution that speaks a separate language, even though it seems to use words in your language. There are forms to fill out with words you don't understand. Your advisor may not make sense, and you don't understand what they're explaining to you. You're afraid to tell people you have no idea what's happening because then they'll find out you don't belong there.

It can feel like all your peers know what's going on and you don't. We don't mean the academics— *that* you're probably doing well at. It's similar to all the other schooling you've done, just more self-directed. It's all the other stuff you struggle with—this gap you can't quite define, but you can feel it all the time. It sometimes seems like there's a conversation going on you can't quite hear through a door you didn't know was there. But everyone else hears it and is taking notes.

Then you go home to visit. Back to your family of origin, perhaps for a holiday break. Your mom asks how school is going, and you start to tell her about your classes because, despite the gap, you're really excited about all the things you're learning. And she looks sort of confused and then asks if you want a sweet cake she made just for you.

The first few times that happens, you let it go. But as time goes on, you get more uncomfortable with telling your mom, or really any of your family, about the things you're learning. It's not that they don't care, but it's like they tune out your play list when you talk about it. So when people ask how school is going, you just start saying "Good! It's great!" and change the subject.

You start to feel like you don't fit at home, which is really weird, you think, because you know these people. If you come from a lower-income family, like Sharon did, you may start to think that people are treating you differently, like you're getting too big for them. And that can be a shocking feeling—you're just trying to get the education everyone said you needed to get.

It starts to feel like you don't fit in your family, and you don't really fit at school. This feeling can really shake up your ethics and sense of self—which may be a good thing because it makes you think about them, rather than taking your installed OS for granted.

You're not wrong, but you're also not right

We can tell you these feelings are true and real and not your imagination. First-generation college students have a disadvantage because they don't come from "the college background." You can't call home and ask your parents what to do because they have no idea either. There's a gap in your OS that students who come from a college background don't have.

That gap is hard to deal with. Students who come from a college background don't realize they have that advantage. The few times you sort of sideways mention how you're struggling with some of this to your fellow students, they glide past it. You learn not to say anything, even to other first-generation students.

Your family gives you feedback that makes you uncomfortable sharing what you're learning. So you just respond in two-word sentences and then change the subject. And get out of there as fast as you can.

When you go home, your family still loves you and is proud of you. But the system you describe and the ideas you're sharing are out of your family's experience. You're changing your OS—you're installing new apps, if you will. And the new apps are changing your entire OS.

Story time

Sharon married when she was 16 years old to a 22-year-old man. He joined the US Army and she became a military wife. She started running her own household just after she turned 16 and had her son just after she turned 17. While she was a military wife, they lived in Georgia, Arizona, and then were sent to the NATO base in Naples, Italy for a few years. To no one's surprise, the marriage ended just after she turned 21.

She returned to her parents' house with a 4-year-old child. She had no high school diploma and no job skills. Growing up, her parents were working-class people with no money to spare. There were no discussions about going to college because people that come from where Sharon comes from didn't do that. Her older brother put himself through two years of community college and had gotten an Associates degree, but that was as far as any of the siblings had gone.

The only job Sharon could get was waiting tables to support herself and her son. And she enrolled in community college. Even today, at this distance, she still can't tell you why she signed up for college, except she needed to support her son and waiting tables wasn't going to provide any stability or a path out of poverty. She had no idea what she wanted to study or any plan after "Take college classes and things will be different."

But she did well in classes. She stumbled into anthropology and fell in love. One instructor suggested she consider working towards a PhD. "You're smart enough to do that." she said. So Sharon applied to a local university for her bachelors and then eventually enrolled in graduate school to pursue a PhD. Today, she still doesn't have a high school diploma.

When Sharon shared what she was studying with her mother, her mother said "But what can you do with a degree in anthropology? I don't know why you're wasting your time with that when you can't even get a job in it."

While that comment seems harsh, and Sharon took it as harsh at the time, with the clearer understanding Sharon has years later, we think we know what her mom meant. Her mother wasn't devaluing the education—rather her mother, with no background in college herself, didn't understand what Sharon was studying or why that was a thing worth studying. To her working-class mother, an education should get you a good job. And her mother wasn't wrong about that, but she wasn't right either.

You decide who you are

During this period, you get to decide who you are. Your parents don't decide who you are. No one but you gets to make that decision. It's up to you. And this can be a hard thing to figure out. And while you're in the thick of it, you can feel out of place in all places you function in.

Sharon started feeling uncomfortable when she was around her family during this time and didn't know why. It seemed her mother didn't want to talk about the things Sharon was learning. While Sharon's mother only finished 9th grade, she was self-educated and Sharon always thought her mother knew everything. The more education Sharon got, the more aware she was that her mother didn't know everything, and this made Sharon uncomfortable. Sharon started to feel she didn't fit in her family of origin.

Part of what's happening in college with all these new experiences you're exposed to is you're growing and changing in a short and intense time period. If you're a first-generation college student, especially if you come from a working-class or working-poor background, your family of origin literally can't understand what you're actually doing. They can't relate to the ideas and intense change you're undergoing because they've never been in that sort of environment.

That's part of why you may feel you don't fit at home. At home, people generally are not in this intense time. They're not trying out different ways of being. And that's fine for them if that's what they want. People are allowed to decide what and how they want to be, just as you're deciding that for yourself.

In fact, some people decide the OS their parents installed is good enough, and they never question it or look at other apps they might want as well, to extend the metaphor. Neither of us understands that point of view, but people get to decide what's right for them. They can live in the way that makes them happy, so long as it doesn't hurt other people. We can't tell them they're wrong or bad or stupid because that's what works for them.

If you're in the middle of this time right now and confused about how all these changes are affecting your ability to function, you may want to avail yourself of the counseling services available on your campus. Both of us have used therapists in the past, and their outside view and support has always been helpful. Of course, even if you're not going through this intense period of change, any time your struggling becomes too uncomfortable, seeing a therapist may be helpful.

Your life story: your present and future

Especially when you come from a non-traditional, non-college background into college, it's common to feel, for a while, like you don't belong in either world. It's common that you feel you need to hide or downplay your life story. It can feel like everyone else in college has a more "normal" life story and yours is too different.

You may wonder if someone like you deserves a college education. It's normal to have these feelings, especially if you come from a non-traditional college background. It's also normal to have these feelings even if you come from a more traditional college background.

But your origin story isn't a life sentence.

Your origin story is important and valuable—it's the story of who you are and where you're from. It's the story of you. If you come from the working-class background that Sharon comes from, your life story includes perseverance and hard work. That's a good story. Your life story doesn't have to limit you. You can be more than was expected of you. The fact that you're in college—or completed college—shows you're already crafting your story to fit you.

Sharon married at 16 because she stopped attending high school at 15 and started running wild. Her parents had no idea how to solve the Sharon problem. Sharon didn't see many options for herself, either. Expectations for her were low from everyone.

If Sharon had accepted the ideas other people had about who she was and what she was capable of, she would never have gone to college. She would never have had the life where she's teaching at a university and writing books. The closest she would have gotten to college is perhaps a janitor or a clerk in food service in a college. But for some reason, she decided she was more than what she was told was possible for her. Her origin story didn't define her life story.

You get to decide who you are and who you want to be. Regardless of where you come from and what the expectations of you are, ultimately, you get to make these decisions for you. And if the options you choose don't fit, you can choose other options. You get to write your life story.

Who you say you are

Out of all this intense college period, you start figuring out who you are. And that's a normal part of being in college. For people who come from a college background, their parents expect and understand the changes because they went through similar changes in their college years. For those who come from a non-college background, your parents may be perplexed about the changes you're experiencing.

As part of this intense college process, you learn new things and are exposed to new ways of being. Sometimes, you try new ways out for a bit to see how they fit. Some you integrate into who you are, and others you discard as not right for you, like a poorly fitting hoodie. You learn more and know more, which may change what you thought were core values for you. Some people in your family may see this as a betrayal of your culture. It is not.

Through this period, you start to feel more solid about who you are and what matters to you. You change the OS to suit who you uniquely are. You integrate these new ways of thinking and new experiences into who you decide you are. You decide what matters to you and what your values are.

And this is good and normal and exciting.

We can tell you that your confusion about how you fit in your family will probably settle down in your mid to late 20s. You will settle into who you say you are and how you fit into your family of origin. Just keep visiting, keep interacting, and trust this is temporary. It most likely is temporary, and you'll reintegrate in your late 20s.

Who you show us to be

You don't decide who you are in a vacuum. We, the other humans on the planet, are also inter-acting with you. We're sitting next to you on public transit, in the movie theater, in classes. We're dating you, employing you, and working on projects together.

Through those interactions, we hear you tell us who you are. We see you under stress, in good moods and bad, when things go well or don't, while you live your life as we interact with you. We get to know you as the person you are.

At first, we accept who you tell us you are. That's how humans seem to be wired. We assume the people we interact with are truthful people. We have a strong cognitive bias towards assuming people tell us the truth.[1] Other people can't see into your head to see who you are—we accept who you say you are. If you say you're a person of loyalty and truthfulness, for example, we accept that at face value. Especially, it turns out, if we know you well.

---

[1] *Talking to Strangers* (Gladwell 2019)

But humans are also watching you. And we're (perhaps unconsciously) watching to see if your actions match who you say you are. If you say you're loyal, for example, do we see actions that indicate loyalty? If you say you're a truthful person, do we see actions that align with our expectations of someone who is truthful? We want to make sense of who you say you are and who you show us you are.

For example, you may have started dating someone. At some point, you talked and decided to be relationship partners. As part of that conversation, you may have discussed what being partners looks like for each of you. Perhaps you mutually agreed monogamy was your preferred partnership structure. Excellent, you both said! Now we're exclusively dating and romantic partners.

And then you may have discovered that the other person has a lot more people in their definition of monogamy than your definition has. It really doesn't matter how much they protest, you know by their actions this is not a person of their word. They said they valued monogamy in a relationship. But their actions showed you that's not at all what they valued. There's a mismatch. A painful mismatch between who they say they are and who their actions showed you they are.

Who you show us to be is your reputation. Your actions map to what you say you will do. Your reputation in the workplace is important if you want to work on the interesting projects that require a high level of trust from your teammates. Your workplace reputation is literally worth money.

More story time
When Sharon was at the end of her junior year as an undergraduate, her father died. It was not unexpected, but it was sudden. Although Sharon and her father didn't get along, it was still a hard thing for her. When a parent you don't get along with dies, it kills any future reconciliation, and that's very hard.

Final exams were a week away, and she didn't feel she was going to do well because she was struggling to focus. She went to her professors and they understood. She took an incomplete in World Prehistory but managed her finals in her other classes.

Fall quarter started, and it was time to clear the incomplete by taking the essay final exam. The night before, she started reviewing the material and confused herself. It was 50,000 years of prehistory. Her study method depended on how the professor was approaching testing.

She called a friend to see if the friend still had the final exam from the spring. Her friend did and ran over with it. Sharon read the final and got insight into how the professor approached testing the material. Now she had a study plan.

The next morning, she got her son off to school and walked to campus. As she walked, she reviewed the material in her head. And then suddenly had a thought. "What if he gives me the same final I looked at last night? I've already seen that final exam…." The more she thought about it, the worse it got. If she took a test she'd already seen, that was cheating. But if she didn't take the exam, she would fail the class.

By the time she got to the building, she knew if he used the previous exam, she couldn't take it. Even though it meant she would fail the class, and she would get kicked out of university.[2] If she got kicked out of the university, she and her son would wind up homeless on the streets.

Luckily, the professor used a different exam than the one Sharon had seen, she wrote the final, passed the class, and she didn't get kicked out of university.

When Sharon tells this story in lecture, she asks for a show of hands from the students who think she was an idiot for deciding she couldn't take the exam if she'd already seen it. She gets about 30% of the room raising their hands.

Sharon realized she was unable to take the exam if she'd already seen it, because she succeeds or fails based on her own efforts. She doesn't cheat to be successful. She works hard and any success she has is because of that hard work. And if that means she sometimes fails, that's also what that means. But she's honest about the work and her efforts and doesn't take shortcuts, like cheating, to get ahead.

And here's the problem

After you figure out who you are, you have to be that person. Who you are when you're engineering is the same person when you're not engineering. You're not two separate people living two separate lives. So if you say you're an honest person, you need to be an honest engineer, too. Your actions in both areas of your life need to align.

And that includes even when no one but you will know. The professor wouldn't have known Sharon had seen that exam if he reused the previous exam. The friend who gave her the old exam would not have known it was the same exam. The only person who would know was Sharon.

---

[2] She wouldn't have been kicked out of the university. She would have had to retake the class. But this was something she had no way of knowing because she didn't understand how universities work. This is typical of first-generation students.

And that mattered to Sharon. If Sharon says she is an honest person who succeeds on her own abilities and doesn't cheat, then that's who she is all the time—even when no one else is watching and no one else would know. Sharon would know, and that is enough. If she got an A in that class because she saw the final exam, she didn't earn that A. That A would forever be on her transcript, but it wouldn't belong to her. It wouldn't reflect her ability and her knowledge.

This alignment is the problem for many people. Many people don't know about or achieve an alignment of ethics, regardless of what they do for a living. They're one ethical person when they're working and a different ethical person when they are not working. They may be loyal and honest in their personal life, but have no difficulty taking bad shortcuts and lying in their work environment. And that's a decision they get to make for who they decide they are.

For example, one of the things Sharon says is true about her is she is a kind person. Not that she doesn't lose her temper, but she tries to approach the world kindly. She tries to allow people to be who they are and not make them wrong if they're different than she is. If that's who she says she is, that also means she can't work in certain industries. That's the logical conclusion of who she is. She can't work in the defense industry where, although it would be super fun to participate in testing things that blow up, the things that blow up are for killing people.

Who you are is the same person all the time

Who you are as a person is who you are as an engineer. In general, how you do one thing is how you do everything.

As an engineer, the rest of society *needs* you to align who you say you are with who your actions show us to be. Bad engineers, unethical engineers, can kill people. It's one of the few professions where you can profoundly change the course of someone's life with what you choose to do or not do. The rest of us count on you and your engineering to improve our lives, not kill us.[3]

When you take shortcuts, when you're careless, when you steal other people's work, when you don't consider the ways your engineering can go wrong, we see you. We know who you are. You show us you are an unethical engineer and an unethical person. That's the person we know. And that person is a danger to the rest of society because we can't count on your engineering.

---

[3] "As engineers, we must consider the ethical implications of our work" (El-Zein 2013)

Don't take your original OS for granted, or as given. Think and examine your personal and professional ethics. Consciously decide who you want to be and then *be* that person.

Be who you say you are, even when it's hard or no one is watching. Make your actions align with who you say you are. You're going to face plenty of ethical situations in your life, both in your career and in your personal life. Sometimes it's clear what you should do, and sometimes it's messy and murky. But you and the people in your life are counting on you to be who you say you are. All the time and every time. We who benefit from your engineering are counting on you to be who you say you are in your engineering, too.

Your reputation is counting on you. Your engineering is counting on you. The rest of society is counting on you and your engineering ethics.

CHAPTER 9

Flow of a Project in a Company

When you're a young-in-the-field engineer, it can seem like projects in your company start and stop randomly and capriciously. While that certainly happens in some companies, most companies are more deliberate than that. But from your position as a young engineer, it can seem crazy. You're working along on the Blue project, it's going well, and you think you'll complete your work in a few months. Suddenly, your boss appears and says, "The Blue project is off. Check in your files. You're on the Green project now, and they have their team meetings in the Pinot Noir room Wednesdays at 9am."

Most frustrating, this is the third time this has happened to you in six months. It seems that you never finish a project before the next one starts. How is the company ever shipping a product to make money? What's going on?

It usually turns out a lot is going on above your head at higher levels in your company. Because what's happening is typically way over your head, you have no idea it's happening; you just see the on-the-ground results. But understanding what's going on above your head helps you contribute in a more meaningful way to the company as a whole. And it makes your life easier when things change this fast.

The rest of this chapter discusses how a project generally flows through a company. While your company may have some differences, the overall flow described here is typical of most companies. This process may have some differences if your company is using an agile development method, but the overall flow, as shown in Figure 9.1, is still accurate.

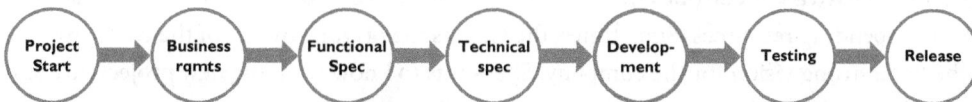

Figure 9.1 – Overall project flow

The titles for roles we're using are the most common for that role or are the generic words for that role. The roles are also not listed in any order of importance because the roles involved have no order of importance in this context.

A note about the roles listed in this chapter: In a smaller company or in an agile environment, these roles are flexible. They are often handled by the same people in different stages of the project or in different phases of the sprint. It's not uncommon for multiple people to take parts of the roles described here as the project moves through different phases. In a large company or a waterfall environment, these roles may, in fact, be different teams or at least different people.

But regardless of whether the company is small or large, the tasks associated with each of these roles must be accomplished to get the product out the door and into the hands of the customers.

Project start

A project typically starts when someone, like a customer, says they love your product, but they really wish it did X. If it did X, the customer would buy it and upgrade all their sites. After asking around, other customers may also want your product to do X.

Or someone has an idea for a new product called Y. Neither situation is enough to get a project off the ground. Every project goes through a process. And doing any particular project means another project may not happen because the resources to do something else are not available.

Companies don't have unlimited resources to expend on projects. Smart companies expend resources on projects with a good return on investment (ROI). What makes a good ROI can vary from company to company, depending on the direction of the company at that point in time. This balance of resources and ROI can be difficult to understand and see in a company. But it can explain why a project everyone thinks is a good idea doesn't happen.

The roles involved at this investigative stage of the project can include:

- **Chief Executive Officer (CEO):** The CEO often has a group of investors to whom she must justify spending resources. Sometimes, the CEO is the original owner of the company, and she has a strong vision for the company. She wants to know what the new projects are and why they are being done, especially if the project is significant to future revenue, is opening a new market, or is otherwise a serious project. If the project is trivial, or fits into what the company is already doing, she may not be involved.
- **Vice President (VP) of Sales:** This person is responsible for the company's sales numbers. He lives and dies by the sales numbers every quarter. Typically, he also reports to the investors. He knows what can be sold to existing customers, who are the best source of income for any company. He knows if Project X can be sold and the length of the sales cycle for this project.

- **Chief Marketing Officer (CMO):** The CMO is responsible for making sure the market wants and needs this product. She is aware of what the competition is doing and what they have coming in development. She knows what the market wants from your products, which may be different from what the market wants overall. She also knows what the pricing should be and how to advertise.

- **Chief Financial Officer (CFO):** The CFO is responsible for the money in the company. Period. That's what she does. She monitors the cash flow in the company and keeps an eye on profitability. She often must justify expenses to the investors and explain why financial decisions were made and why they were made the way they were. She is also responsible for cost controls and can kill a project if she thinks it costs too much.

- **Chief Technical Officer (CTO):** The CTO is responsible for the technical big picture of the company and what the technical possibilities of the company are. She knows what the code base allows, for example, and what it won't, both today and in the future. She may have oversight of manufacturing if you're a hardware company.

At this stage of the project, these roles discuss the marketability, the money spent vs. the potential income, and the technical requirements for this project.

Many projects die here because they were good ideas but:

- cost too much to do,
- don't have the potential market size to make it worth the effort,
- a competitor is doing something similar,
- are technically too hard for the company to implement at this time,
- don't fit with the direction the company wants to go at this time.

And sometimes a project dies here because it's not what the CEO had in mind when she founded the company. Sometimes a project lives here because the CEO loves the idea even though it may not make a lot of financial sense. Maybe she's in love with the technical challenge of the idea.

The deliverable from this phase can be as informal as an email that says "Sure, scope it out." That doesn't mean the project has been given a green light, but it does mean permission has been granted to go to the next phase.

Business requirements

The next phase of this project is the business requirements, where it's determined what's needed for this project to create a product that will do what it's supposed to do. Think of this phase as defining the problem space and how the problem might be solved from the user's point of view. Sometimes in this phase, you work with the customer that requested X to validate the requirements and the high-level solution with them. No one is looking yet at the specific details of how you're going to solve the problem. How you're going to solve the problem is the next phase.

The roles involved at the business requirements phase of the project can include the following:

- **CEO:** In a small company, the CEO is often more hands-on and wants to know what's being developed. Frequently she was the original architect of the products and she understands the problem space because she lived it. In a large company, the CEO is often invited to the meetings, as a courtesy, to let her know what's going on with that project, especially if the project is groundbreaking, opening a new market, or otherwise high profile. She doesn't like surprises, regardless of the size of the company.
- **VP of Sales:** The VP is interested in what problem is being solved because she knows other customers have this problem. Other customers with this problem are potential customers for this product. She knows how the sales team will need to talk about it, too.
- **CMO:** The CMO wants to know how the project fits into the rest of the marketing plan and/or what she needs to do to market this project. She is involved in setting the price for the project and knows what the competition is doing that may be like this. She also knows what the size of the potential market is for this product.
- **CFO:** The CFO wants to know how big the problem is and what it will cost to solve it. He may cancel the project at this point if the scope/cost of the problem is too big. Or maybe the cost is reasonable but, based on the size of the market identified by the CMO, the return on investment is too small. For example, it makes no sense to pay $50 million to create a product if the total market value is $40 million. Perhaps that money should be spent elsewhere.
- **CTO:** The CTO knows the architecture of the current products and what can technically be done with them. She also knows the technical debt the company has accrued in past projects and how that impacts future projects. She knows what is needed from the current teams to do the project. She can kill a project if the technology required for the project is beyond what the company can handle at this time.

- **Human resources (HR):** This is a new player in our discussion. HR should be involved at this stage of product development because you may need additional staff. Perhaps you don't have people you can move off projects, or perhaps, after the CTO weighs in, you discover you need to use technology the company doesn't have experience with. Maybe you need to hire an outsourcing company to do some or all of the development. You need HR to know what you're thinking about so they can inform you if any of these solutions will work. For example, you may need to know whether there are roles available on the job market who have the skills needed to do this project.

These roles work out what the problem is, how you might solve it, how big the market is for a solution, what you can charge for a solution, what you need technically to solve this problem, and who you may need to create a solution.

The deliverable from this phase is a document called *requirements*. Some companies call it *business requirements*. If the project makes it out of this phase, it doesn't mean the project is going to go into development yet; it simply means there were no business reasons to stop it here.

Functional and technical specification phases

If the project makes it through the requirements phase, it gets into the *functional specification* phase. This phase is where you define the scope of what you are going to build. Many companies skimp on this phase of the project, especially if they are working in an agile environment. *Don't*. Ignoring this step always results in a product delivered late without the features that are needed for a minimum viable product.

In a hardware environment, this phase must happen because at some point, you need a factory to make this product. This means you need to define what it does at a high level so you know at least what kind of factory you need to work with.

If you're working in an agile environment, this phase may be called something else, like *generating epics*, for example, but this phase does happen. It may be very informal, but it must happen or you'll never have a strong idea of what you're building. If you don't collectively have a strong idea of the product you need to build, you'll each build something a little different. You all need to understand what you're building if you're going to eventually ship it to customers. And if you don't ship to customers, you won't have jobs for long.

The roles involved in the functional specification phase can include:

- **Sales:** In this phase, you're no longer dealing with the upper management of sales. At this point, you probably have what's known as technical sales involved. The technical sales people are the backup roles who help the sale along. They often do demos and answer technical questions for the sales people. They know what customers ask about and what problems they have. They also have insight into where this product fits from a customer perspective.

- **Marketing:** As with sales, this is not upper-management marketing. In many companies, this is the technical marketing group for many of the same reasons the technical sales people are involved. The technical marketing group knows the market for this problem space, what the product needs to do, and how it compares with competing products.

- **Architect:** At this phase, the architect knows the current architecture for this product line or technology area. She knows how it works and what can be done with it, and she also knows if an entirely new architecture needs to be developed for this project.

- **Development:** Now, finally, you have the development team involved. This is usually a high-level development manager, typically from the group that will make the product. She knows what her people can do, who should be on this project, who is going to be out on parental leave, who is looking for a new project because he's bored, and so on. She has hands-on knowledge of the roles that need to be filled to make the product.

- **Project management:** The project manager is responsible for defining the resources needed for a project. He knows what other projects the company has going, the status of these projects, and their priorities. He knows how to run a project from start to finish in the time frame required. He can tell you what resources are needed to get this done. He can also assign time/effort to each feature or phase, so you can decide what can be delayed for this version.

- **Product management:** The product manager is responsible for this product or similar product lines. She has a deep understanding of this problem space, and she knows how this product supports this space. She also knows how the other products in this product line support this problem space and what the differentiators between the products are.

- **Quality assurance (QA):** Quality assurance is responsible for testing this product to make sure it works and that it works as designed. Why is he involved so early? You have nothing to test yet. But you will have a product to test at some point, and he needs to work on the testing plan. He needs to know what you are building and how it should work so he can determine how to test it. Often, a really good QA person can help define the project by watching the variables that need to be tested. He can also give thoughts on previous, similar products that were impossible to test or never passed their test cases.

- **Technical communication/UX:** The technical communication team (sometimes called user experience (UX) writers or content designers) needs to start planning to write user-assistance content for the product. This includes content embedded in the product, such as assistive text in text fields and user interface (UI) content, as well as manuals and other content for users.
- **Factory:** If you're building a physical product, you need a factory to make it. Most companies don't own a factory—it's cheaper to use a factory someone else owns. Even Apple uses an outside company to build many of its products. This can make good financial sense because factories are expensive to own and operate. Where the factory is located in the world drives what you pay for labor and often materials. Paying more for these costs means selling your products for more or taking a smaller markup. A better business decision may be to use an outside vendor and make the CFO happy. But you need to make sure the factory you want to work with can build what you need and has manufacturing capacity available when you require it. Are they going to be buried in Christmas orders when you require them to start your product? Do they need to hire different staff for your technology? During the pandemic, we all learned of supply chain issues at the ports. Maybe paying more for a closer factory is better? Better to know this now while you still have options.
- **IT or web support:** If you're delivering the product on the web, or if it's a cloud-based product, you need an IT web-hosting person to be involved. She can tell you if she requires more servers for the required uptime. Maybe she knows the current web architecture won't allow the planned number of users, and it's time for a server farm. Maybe the existing virtual server setup can't manage the planned technology and altogether different virtual servers must be brought online. Or perhaps, with a small change to the planned technology, you can deploy with the technology you have.
- **Materials:** If you're building a physical product, you may need the materials engineers involved. It could be that, with a new material, you can reduce costs or increase reliability. Perhaps the materials you think you should use are expensive and another, similar material can suit the purpose. Regardless, she can tell you what you need and how to get it. She can prevent you from making mistakes that would make this project more difficult.

The deliverable from this phase of the project is called the *functional specification*. In some companies, the functional spec and the technical spec are rolled into one document, in which case, both together may be called the functional spec (see the section titled "The nature of a functional specification" in Chapter 16).

At the end of this phase, regardless of what you call the deliverable, you have a document that defines, at least at a high level, what the product is and how it's going to work. If the company does a technical specification, the project moves to the next phase. The technical specification is more common in a waterfall environment.

> The technical specification defines the low-level details of the features and functions described in the functional spec. For example, for a software product, the functional spec might say the product needs a database. The technical spec would define the database structure and how the various fields function. For a hardware product, the functional spec might say the case must be robust. The tech spec would define the exact material requirements, such as the tensile strength required for the case.

Development (and testing) phase

Finally, it's time to start actually making this product. If you're a young engineer, this is when your boss tells you you're now on the Green project. They meet in the Pinot Noir room on Wednesdays at 9am, starting next week. Talk to Joan because she's your team lead. Finish up your tasks on the Blue project and get your code checked in.

In this phase, team members or roles may ebb and flow in their interests, appearing and vanishing as the project goes on. This is normal and not necessarily because they lack interest in the project. At this phase, many team members are involved in multiple projects in different stages.

The development and testing phases overlap and are interactive. Because these phases overlap so much, we deal with them as one for ease of description here. Regardless of whether you use a waterfall or agile development methodology, as parts of the product are completed, they go into testing. As they are tested and bugs are found, they roll back into development for further bug fixing and then go back again to testing. Agile typically does this faster than waterfall.

If you're in an agile environment, this phase may run in a two-week sprint, often delivering the results of the sprint to customers. However, not all sprint results are delivered right away. Sometimes a larger project is broken into sprints for ease of development. It depends on what the project is, what the deliverable is, how big the product is, and other variables.

Regardless of agile or waterfall, the roles involved in the development and testing phase can include:

- **Architect:** In this phase, the architect is more of an advisor to the product. She may attend meetings and/or stand-ups, but may only be available for thorny technical questions.
- **Development:** This is the group of people actually developing the product. It's usually your team, but can include other roles, such as outside development groups and contractors. This team spends the day working through the tasks needed to make the product.
- **User interface (UX) designers:** If you're creating a product that has a front end that people need to interact with, you may have this new team member in this phase. They may have advised in the technical specification phase if your company does that phase. Their job is to make sure the user interface is usable, which probably doesn't mean it's beautiful. Some UIs are also beautiful, but well-designed does not necessarily mean it's pretty. UX design is an important part of a usable product. Poor UX design can kill a product.
- **Project management:** Project management keeps the project on track and makes sure you have what you need to hit the milestones in the project plan. At this phase, if your company is using an agile development method, this role can also be a scrum leader. Some engineers think the project manager or scrum leader is there to be a pain in the neck. As a result, they sometimes hide problems from the project manager or scrum leader. This is a bad idea. A good project manager or scrum leader wants to know about the issues so she can resolve them and keep the project on track. If she doesn't know about an issue, she can't solve it. Don't hide things from your project manager or scrum leader.
- **Product management:** Your product manager is interested in making sure the project solves the problem correctly and meets the needs of his users. Because he knows the problem space so well, he knows what makes sense to the users. A good product manager may also want to do private demos to select customers to get feedback while the product is in the development phase. Support this effort, even though it may cause you to rework things. The product will be the better for these early alpha reviews.
- **QA:** QA is tightly involved in the development and testing phase because they are often finishing test plans and want to know when things change. They are also concerned about when part of the product can go into testing or be re-tested. When parts of the product are ready for testing, they use their test cases to test the product. They frequently report about their testing in the development-phase meetings. In our experience, testing is one of the most frequent delays in a project. Testing takes time, and there is often nothing you can do about that.

- **Technical communication/UX writers:** No product can ship without instructions. They may not be in a printed book, but at least instructions are embedded in the UI. The technical communication team creates this content. In some companies, the content deliverable is a help system and in others the technical communication team does everything from product UI text to creating training for end users and support. This varies from company to company. While the technical communication group is developing content, they are also usually working with the product, so they do some informal QA. They examine things like task flow and can give advice about the usability of the product. If you're translating the product into other languages, they are also involved in the localization process. More about localization below.

- **Factory:** During the development process, the team is communicating with the factory, getting prototypes, and generally determining if the product you think you're building is the product the factory is actually building. You would be shocked at the differences we've seen, including forgetting that the ink jet cartridge in a printer needs to be accessible so it can be changed. The prototype didn't have the cartridge located where it could be replaced, limiting the amount of printing this $700 device could handle to one cartridge.

- **IT or web support:** If you're building a cloud-based or other web deliverable, the team that is responsible for eventual deployment is involved to run tests, making sure the infrastructure they think they need is the one they need. They can also tell the team when something can't work because of infrastructure limitations and suggest workarounds.

- **Training:** If your product is one that requires training for the customers to use, the training group develops the training guides or courses. Your company may deliver the training as an online set of courses or as in-person training, but either way, it needs to be developed for the eventual end users. Like technical communication content, training content needs to be accurate and delivered on schedule.

- **Professional services:** Some products require customization or configuration to match a particular customer's needs. Professional services is the organization that typically handles this task. They get involved in this phase to begin learning how to configure or customize the product and to influence decisions that affect them. Because they are looking over the entire product, they can identify issues between development teams that impact the entire product.

- **Localization:** In this phase, the localization team is translating UI text strings and putting them into the product as the strings get completed. While the localization team is not really testing in the development phase, they can often identify issues they find as they work with the product. Very often, they are localizing and testing at the same time.

- **Support:** Although there is no product yet, support is looking for what is needed to provide support for this product. They may have suggestions to improve the product to reduce the support burden, especially if your company has developed similar products in the past. Frequently, they can forestall issues that will cause user problems when the product is released.

The development and testing phase doesn't really have a document deliverable, although many documents are created and updated in this phase. Both specifications are updated as the details change, customer stories are updated, Jira tickets are created, and test suites are developed and updated. While it seems like this phase might be a Wild West while you all work like crazy to get the product done, it is usually a pretty tightly run phase, certainly for experienced companies. A good project manager or scrum leader can make or break this phase.

Reality check meeting

At some point in the development and testing phase, typically 66% to 75% of the project schedule, it's time to have the *reality check* meeting. At this point in the schedule, it becomes obvious you cannot deliver the product as specified and something has to go to meet the deadline.

We are constantly amazed that companies don't scope projects better. Humans have been developing software, for example, for over 50 years, and teams still can't accurately scope what can be done in the time available. They regularly overestimate by 25% to 33%. One of the things that the agile method was supposed to fix was this inconsistency in project scoping and yet, here we are.

The reality check meeting involves most or all of the teams identified in the development and testing phase. You know this is happening because high-level managers from the teams appear and meet in the large meeting room. The door is closed and intense discussions start. Voices may be raised, depending on the culture of the company. We've seen these meetings go for several hours or as long as several days.

In a functional company, people passionately argue for the part of the product they are responsible for. Someone's hard work is going to die, and no one wants it to be their work. They are proud of it, and they want it to ship with the product. But resources are limited, time is finite, and the product must ship. Something has to go. No one wants it to be their thing that goes.

Eventually, decisions are made and agreement is reached. The minimum viable product has been identified. Sometimes people in the room don't like the decision, but consensus (or the highest-ranking manager) rules. Sometimes people are really mad at the decisions made.

A result of this meeting may be that your boss says you're off the Green project and moved to the Orange team to get another feature finished. If you don't know about the reality check meeting, it can look like projects start and stop randomly—which does happen in some companies—but decisions are happening over your head. The scope of the product is reduced to what can be delivered in the time available with the roles (resources) available to do the work. The product is still in the development and testing phase, but with some features removed for this version.

Delivery phase

The *delivery phase* is where you finally deliver the product to the users. All the work that everyone has put in finally results in something you can ship. Based on the reality check meeting, what you're delivering may be less than what was originally specified, but it satisfies the requirements for a minimum viable product plus a little more.

Additionally, when you get to this phase, you may know there are parts that your team jammed together expecting that you will go back and fix them in the next release. This is called *technical debt*—things that are not really done properly but were slammed together to finish the product. Never let perfect get in the way of done, unless you're working in a tightly regulated field where lives depend on your product working perfectly.

The problem with technical debt is that it gets added to the next iteration of this project, adding to the tasks that need to be done to get the next version complete. Technical debt can be part of why what was put in the specification had to be reduced. In some companies, at some point, a release may actually be solely for cleaning up the existing debt. You've reached the point where maintaining the code costs more than rewriting the code. Don't be surprised if you're put on a project that appears to be nothing more than a rewrite. These are expensive and critical projects.

The build process is usually all hands on deck. People are often in the office for several days while the last issues are dealt with as part of the process. You may sleep at the office until this is finished. You almost certainly will work 70 hours a week to get the product out the door. This is called *crunch time*. If your employer thinks every day is crunch time, it's time to consider whether you learned all you can from this job and need to find a new one.

In this phase, you hear some new phrases, including *gold master*, *general availability*, *GA*, and *gold* (also, probably, some new words). All these refer to the final "thing" that you're releasing, regardless of whether it's a software product, a hardware product, or a hybrid.

Depending on the development process you're using, different roles may be involved in this phase, but as a rule, the teams listed below are part of the process.

- **Build master:** New to this phase is the build master. The build master is responsible for making sure the code is checked in, compiles as expected, and can be delivered as planned. This is all the build master does—it's her responsibility to make sure this happens. If you don't follow the best practices of your company, for example, you can break the build. Breaking the build is a bad thing because it prevents the build from successfully completing so it can be delivered. If your code wasn't checked in properly or wasn't named correctly, and you broke the build 22 hours into a 30-hour build, the build master will be mad. You'll know you broke the build because the build master will appear breathing fire at you. You have no defense when she shows up because she knows who broke her build. It was you. Admit it, fix it, apologize, and don't do it again.

- **Sales:** Sales is back because they want to know what and how the product works because they need to get ready to sell this to customers. They have emails to write, conversations to pre-script, and technical sales has demos to put together. When it's time to ship, they need to be ready. Dysfunctional companies often don't include sales until this point. If you find yourself working for one of these companies, get your résumé ready—it's only a matter of time.

- **Development:** Development is on call during this phase because bugs are going to be found at the last minute during the step called Smoke Test. Smoke tests involve putting the product through some basic tests that indicate the product is generally functional. The smoke tests are going to fail because that's what happens. Development needs to be right there to find the issue and solve it.

- **Marketing:** Marketing is involved again because they need to make sure they understand what the product actually does, how it fits in with other products, and what specific problems this product solves. They need to create data sheets, marketing literature, probably email and web campaigns, and other marketing content. Again, in a dysfunctional company this may be the time Marketing gets included, and that's honestly just too late. Run while you still can.

- **Project management:** Project management is tracking the details that make up a delivered project. They have moved the features that died in the reality check meeting to the structure for the next version of the project. They are making sure their project plan includes all the last tasks and adding the time values to make sure all the parts are accounted for. In an agile environment, the scrum leader is putting together a retrospective that details how this project or sprint went and lessons learned for next time. The history of this project is critical for future projects, and these details need to be accounted for while they are fresh.

- **Product Management:** Product management is here to make sure the product is what the users need. Product managers need to be involved in this phase of the project to make sure they understand how, in detail, the product works and how it fits with all of the products in their portfolio. If parts of the product aren't what they were hoping for, they make notes for what to fix in the next version.

- **QA:** In this phase, QA is working overtime to make sure last-minute bugs found in the smoke test are fixed. At times, small bugs may be allowed to ship with a promise to fix them in a patch. But if you're working in a hardware environment, these bugs sometimes must be fixed before the product ships because patching hardware after delivery can be difficult.

- **Technical Communication:** Technical communication is finishing up any user-facing documentation, writing release notes, and documenting any bugs that are shipping with the product, along with any workarounds for those bugs. They may also be creating training manuals/materials, depending on the type of product. If these materials are localized, they are also making sure the content is ready to be translated.

- **Support:** Support is involved to know what the product does and how to support it. They may be writing knowledge base content, creating technical release notes for internal support people, and making sure staff are ready to support the product as soon as customers have it.

- **Factory:** At this point, the factory is waiting for the final approval to start building the product. For hardware products, after the factory starts manufacturing the product for delivery to customers, it's hard and very expensive to stop and re-tool to fix mistakes. No one wants that to happen. Therefore, during this phase, the factory may be making final prototypes for testing and integrating updated firmware.

- **IT or web support:** If your product is a software cloud product, the IT people are doing final tests in live environments to make sure what they think works in fact works. They are running load tests and integration testing. When they get the approval to go, they need to be confident that what is supposed to work actually works.

- **Training:** Any customer training needs to be ready to go. At this point, the training team is finishing the training materials, testing the laboratory exercises, and making sure the materials accurately reflect the product. If the training will be delivered online, they are making sure the systems are working and scalable.

- **Professional services:** The team responsible for installing and configuring at customer sites must verify that the installation and configuration documents reflect how the product works. They also work with QA, support, and tech comm to ensure every detail is accounted for.

- **Localization:** Localization is making sure all the text in the application works correctly in all languages you're shipping in. Some companies ship non-English versions a few weeks later than the English release to make sure the localized product is the same as the English-language product. Simultaneous releases in multiple languages can be challenging, especially in an agile environment. Making sure right-to-left languages in the UI work correctly in an environment where the product isn't firmed up until moments before it's released can be challenging.
- **Shipping:** If your product must be shipped, the shipping department makes sure the product is sent to the places where it needs to go. Sometimes they are shipping it from a warehouse, and sometimes they are fulfilling orders that are sent by a drop-ship company. Either way, they need part numbers in their shipping system, prices correctly entered, and so on, so they're ready when the approval to go happens.

The secret

Every single position covered in this chapter can be, and often is, an engineering position, in that having an engineering degree is important if not actually required. Many engineering programs tell you that your engineering employment options are what we call build-it engineers or teach-it engineers. The programs are not wrong—from their point of view, these are the options—but the division is not that simple.

In the business world, engineering practitioners can and do fill many roles as part of creating a product and getting it out the door. Every role listed in this chapter is important to getting the product built and into the customer's hands. Even the CFO can be an engineer. To perform in the CFO position, you need an additional degree in accounting. But engineers who understand money and how to manage it in a corporation can essentially write their own ticket in the business world. There are not a lot of them out there, and the value they add is significant.

Our point is you may be thinking that writing code all day in a dimly lit cube is starting to sound like a circle of hell. Maybe you like the design bits (as both of us do) but find that building bits is not at all interesting (as both of us don't). Stay with your program and get your degree. You have many options out here if you have the degree. Without the degree, many of these options are not available to you.

CHAPTER 10

Pitching Ideas

During the course of your career, you will almost certainly pitch multiple ideas. You may find yourself doing formal pitches with code/hardware samples, PowerPoint slides, research reports, or a full-on business case. Or you may find yourself doing more informal pitches at the water cooler or as a side issue in a regular meeting. Regardless, know that when you pitch an idea to investors or to higher-ups at your employer, you are writing a business case.

There are three key contexts to understand for a business case:

- **Finance:** A general grasp of what money is and how it flows through a business.
- **Historical/technical:** An understanding of the problem you're trying to solve and the prior art necessary to solve it.
- **Market (size and maturity):** Where your product (or proposed product) falls on the technology adoption curve.

Finance context: how the flow of money fits

We talked about this already in Chapter 5, *The Business Context of Communication*. And we'll talk more at the end of this chapter about how to write a business case for your pitch. But before we get there, let's examine some foundational thoughts.

The key to using what you know about finance in a business case is to make sure that you show that your idea can either make a profit or reduce costs, thus increasing profit.

In any business case, you need to show the following:

- The **cost of doing nothing**. For example:
  - ► Missed potential revenue for a new product or service
  - ► Lower productivity and all the costs associated with that (release delays, unnecessary work, unnecessarily difficult work that takes longer, and so on)
  - ► Legal risk and all the costs associated with that, such as liability from customer lawsuits

- The **cost of doing the proposed thing**. For example:
  - ► Cost to purchase or license new software
  - ► Cost to train staff on new software and/or new procedures
  - ► Cost to purchase different parts or materials
  - ► Cost to adjust the factory tooling to incorporate those parts or materials into manufacturing
- The difference, or gap, between these two costs
- When the company will see the new profits or reduced costs

Without *all* of these elements, you do not have a complete pitch or business case that convinces finance and/or leadership to take a chance on your idea. Many good ideas die before they can be implemented because the engineers proposing them do not show how the ideas improve the company's finances. Don't let your idea be one of them.

Historical/technical context

Prior art and problem-solving

> *If I have seen farther, it is because I stood on the shoulders of giants.*
> —*Sir Isaac Newton, circa 1675, private letter to Robert Hooke*

Prior art, in engineering, refers to all the engineering and/or scientific discovery that has come before, particularly those breakthroughs upon which *your* idea is based. And yes, your idea is based on what has come before. Even as big (shall we say, disruptive) a change as the smartphone could not have happened without, at a minimum:

- telephones
- the transistor
- plastics
- radio broadcasting
- batteries
- the control of electricity
- the steam engine
- wind/water mills
- the pointy stick
- the control of fire

We could go on. And most of the items listed above have prior art *they* depend on. The point is, innovation that matures beyond scientific curiosity into usable, sellable products that improve people's lives depends on prior art.

For example, Leonardo da Vinci designed flying machines of great grace and elegance. He even successfully tested several of them. So why didn't we have flight much earlier? Because da Vinci's inventions depended on human power, which is insufficient to the task (turns out that birds don't necessarily count as "prior art").

Lacking the prior art of a stronger—and tireless—power source, his designs remained impractical for broad adoption. Add the internal combustion engine to da Vinci's understanding of wing aerodynamics, plus a relatively cheap and easy source of powerful fuel, and you eventually get a population that takes public air travel for granted and complains about getting only one tiny bag of peanuts. But without any one of these, we're walking or riding animals and still complaining about getting only one tiny pack of peanuts.

Products solve problems

We've spoken of one aspect of the context of communications in business: finance. While that is a foundational piece of context, it is not the only contextual frame to consider. *Products solve problems.* They may be low-stakes problems ("I'm bored") or high-stakes problems ("I need to accurately identify cancer cells in a CAT-scan image at scale").

Throughout this book, in particular as we look at the phases of the product-creation process, we talk about how to effectively communicate when the context is solving customer problems.

Particularly, for any given inventor or company, products sit firmly at the intersection of People Street and Prior Art Way (see Figure 10.1).

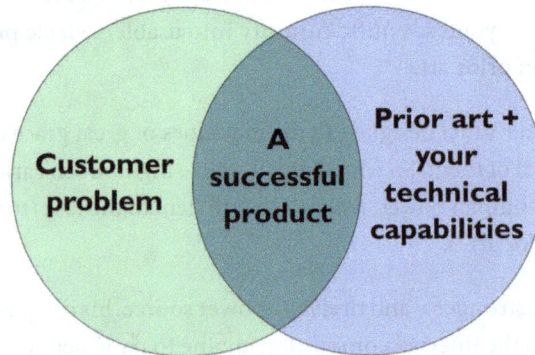

Figure 10.1 – Intersection of customer problem and prior art + your capabilities

In other words, we build successful products using current knowledge (prior art), targeted research and development (innovation), or a combination of the two. Successful products solve problems people will pay to solve. This combination is how we have even a hope of getting any revenue from our finance context.

Makes sense, right? We've got numbers doing repeatable and reliable "numbers things" in our financial context. We've got technology consistently building on itself in our historical context and we've got engineering creativity discovering solutions to problems in our technical context. But the next step of context is where we as engineers start to get uncomfortable. Because you know who has problems? *People.* And people make a market.

Market context: the technology adoption curve

One of the curious things about humans is that we want new developments and innovations—but not too new. Many of your potential customers may not be ready for your amazing new idea. How do we know that? Geoffrey Moore's Technology Adoption Curve.[1]

Although we like to believe that when we create new, outstanding, even disruptive technology *everyone* on the planet will run right out and buy it *now*, that just isn't so. In fact, how people adopt brand-new technology follows the predictable curve shown in Figure 10.2.

---

[1] *Crossing the Chasm* (Moore 2014)

The Chasm

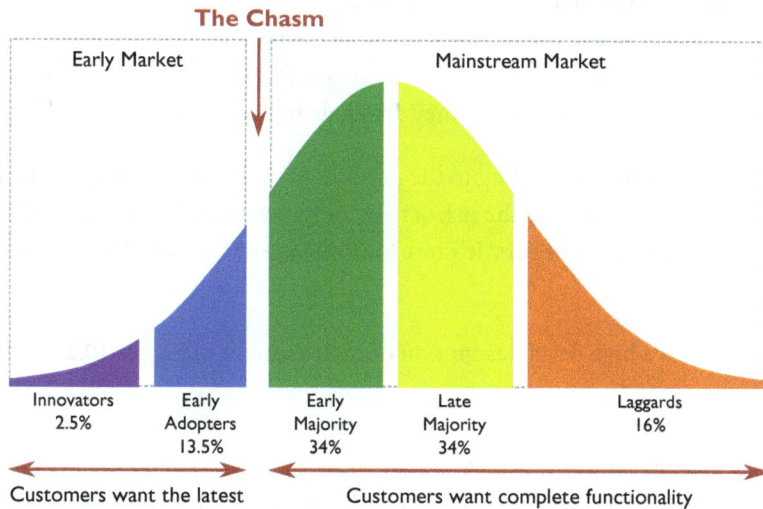

Figure 10.2 – Technology adoption curve[2]

Before we discuss the adoption curve diagram in detail, let's define two key types of innovation: disruptive and sustaining. From Wikipedia:

> … A *disruptive innovation* creates a new market and value network or enters at the bottom of an existing market and eventually displaces established market-leading firms, products, and alliances. The concept was developed by the American academic Clayton Christensen and his collaborators beginning in 1995.
>
> … A *sustaining innovation* does not significantly affect existing markets. It may be either: *evolutionary*: an innovation that improves a product in an existing market … [or] *revolutionary*: an innovation that is unexpected, but nevertheless does not affect existing markets.
>
> —(Wikipedia 2022), © Wikipedia CC BY-SA 3.0.

---

[2] Based on the work of Everett M. Rogers, *Diffusion of Innovations* (Rogers 2003), and later refined by Geoffrey Moore to include the idea of a "chasm," *Crossing the Chasm* (Moore 2014).

In general, your initial target market for disruptive technology falls in the innovators and early adopters slices, while sustaining innovations appeal to the remainder of the chart. Where people and companies fall on this chart has to do with their comfort level regarding risk. The more comfortable they are with risk, the earlier they are likely to fall on the chart.

Placement on the chart varies according to the product as well. For example, some people are very confident and comfortable with the risk of new computer software, but not at all comfortable with the risk of new cell phones or cars. It's rare for someone to fall completely into one category or another for all things.

The following provides a high-level description of each segment in Figure 10.2.

Innovators

Innovators want new, shiny, exciting technology. They are motivated by "Be the first on your block to…," particularly if it increases their tech cred. Or they're just inherently very interested in the tech and like to have the latest to play with. Often engineers, scientists, or inventors themselves, they are tolerant of poorly behaving, difficult-to-use, or buggy systems. They tend to be very comfortable with disruptive innovation and risk.

Early adopters

Early adopters have an overwhelming need to solve the problem the new innovation solves. Their need is usually so strong that they overlook poorly behaving, difficult-to-use, or buggy systems. They complain about them, but they'll tolerate the issues—at least until a different product or solution that is easier to use or more reliable appears.

They are marginally comfortable with disruptive technology, but only if it solves a painful problem for them. In general, they do not tend to be inherently accepting of risk unless the pain of the problem is greater than the pain of the solution. Many of them may also think new tech is just cool or are also inherently interested in the tech.

The chasm

The chasm is the divide between the early adopters, who are comfortable with risk, and the early majority/pragmatists, who are willing to wait for an easy-to-use solution because either their pain or their risk tolerance is lower.

However, just on the other side of the chasm, the bulk of the potential market awaits.

Many disruptive innovations fail to cross this chasm, largely because all the sales and marketing that convinced innovators and early adopters does not convince pragmatists. If you're trying to get your product to cross the chasm, you must stop doing most of the marketing things that seemed like they were working before and start doing new marketing things. Many companies cannot do this.

For example, marketing for innovators and early adopters can focus on new features or even how exciting the innovation itself is. Innovators are not necessarily focused on what's in it for them, because all the things developers who market their products like to focus on (such as cool new features) *are* what's in it for them. For early adopters, their pain is so obvious to them that they are very clear on what they're looking for as a solution. You typically don't need to explain their pain to them or how your product solves that pain. They can see it for themselves.

By contrast, early majority and the members of the rest of the curve usually aren't looking for cool new features. They want stability, comfort, security, ease of use—things many developers don't even focus on in their products, let alone expose in their marketing. So all the marketing that worked to get you to the chasm often won't get you over the chasm. Because many developers are in the innovator or early adopter categories themselves with regard to their attitudes about new technology, they often don't think about the other segments or understand what those segments want. They need to think differently about their products and their marketing.

Early majority/pragmatists

Like early adopters, pragmatists are trying to solve a problem. Unlike early adopters, the problem is not particularly painful for them, or they are not in a position to take the risk. For example, a corporation that must roll out a solution to hundreds or thousands of users can't do that with buggy technology. It is too expensive to support hundreds of users with buggy technology.

Pragmatists tolerate a certain amount of poor behavior, user-interface issues, and bugs, but not for core functions—those must be reliable and easy to use. They are usually not comfortable with disruptive technology, preferring to wait until at least one sustaining innovation cycle has occurred.

Late majority/conservatives

Technological conservatives are suspicious of new innovation and are highly risk-averse. Either they are not experiencing the problem the technology solves, or they believe that keeping their existing methods is less painful than adopting the innovation.

These users have virtually no tolerance for poor behavior, user-interface issues, and bugs. Even nice-to-have features must work reliably and easily. Technological conservatives often initially reject disruptive technology entirely, preferring to wait until several cycles of sustaining innovation have occurred.

Skeptics/laggards

Laggards adopt new technology only when their previous solutions become unavailable. Regarding risk, laggards make late majority/conservatives look like wild technology radicals. Laggards have nearly zero risk tolerance. The pain of adopting new technology is and always will be greater than the pain of their current solution.

These users have no tolerance for poorly behaving, difficult-to-use, or buggy systems—if they encounter difficulty, they abandon the new solution and return to their previous methods or, if they can't return to the previous solution, abandon solving the problem entirely. These users adopt a new innovation only after many cycles of sustaining innovation have occurred, unless their previous solution becomes unavailable or is broken beyond repair.

Bear in mind that this curve applies to *new* customers; that is, customers buying the product for the very first time, instead of updating or replacing an existing instance of the same product.[3] After you've sold to all the possible new customers, your product may become a commodity, as opposed to a disruptive or sustaining innovation, and different rules apply. This text does not discuss the commodity aspects of a product lifecycle.

Communicating in the market context of the technology adoption curve

Understanding these aspects of innovation can help you estimate the size of your market, which directly affects your revenue. For example, if there are significantly more pragmatists than there are innovators, you can sell more units after pragmatists adopt the technology. However, innovators and early adopters will typically pay more for a product that is very technologically "cool" or solves a particularly painful problem. And if the product is truly disruptive, you cannot skip straight to the pragmatists because they need the other segments to validate the product before they will buy it.

---

[3] Think the iPhone 16 (which laggards might purchase) vs iPhone 1 (which laggards might never purchase). Essentially, in 2024, laggards miss their flip phones with no smartphone technology. Some laggards actually miss land-line phones.

Further, how you talk to the groups on either side of the chasm changes. Innovators and early adopters either already know the technical jargon or recognize that they must learn it to be successful. They understand that sometimes new technology gets too far ahead of prior art and ignores some basic aspects of human cognition (much more on that coming soon).

On the other side of the chasm, you must choose your words more carefully to suit audience needs: simpler phrases, less jargon, more explanations, and so on. Pragmatists, conservatives, and laggards not only don't know the jargon, they don't *want* to know the jargon. They want something like what they already have, just…better. To jump the chasm successfully, you need to be aware that the people on each side of the chasm need and want very different things.

> Your co-workers and individuals in leadership may appear in any phase of this curve. In fact your entire company "personality" may fall in any phase of the curve.

Ultimately, you can pick one or more of these segments as the ideal audience for your pitch. You just need to understand the tradeoffs and impact on the flow of money that choice will have.

Practical application: the business case

Mom and Dad probably just take your word for it that your idea is a fabulous one and will make you into the next Bill Gates or Steve Jobs.[4] Probably tomorrow. Or at least by the end of the week.

However, your boss or investors are more cynical: they've heard it all before (and said some of it), and they don't trust you. They worked hard for their company's money and are inherently suspicious of anyone who wants to spend it. They need you to prove that your idea will make them more money than you are asking them to spend on it (in other words, they want a profit). Remember the flow of money—they're not being mean, they're being responsible.

The business case is how you prove the monetary, tangible value of your idea.

In a business case, you ask someone else to give you money to do something, like create a product or buy new equipment. You may be asking your company leadership, or you may be asking a venture capital firm. Either way, it's not enough to just have an idea for cool technology, disruptive or otherwise, or want to have the latest, fastest computer to work on. You have to show that you

---

[4] Unless you're Bonni's kids—she made them write business cases to get their first cell phones.

understand business and how this request will make or save money. That means, in your presentation or business case document, you have to include information about:

- **The market that may purchase this:** Consider the technology adoption curve, as well as further audience information we'll talk about in Chapter 14, *Constructing Explanations*.
- **How this product will make more money than it costs to build it:** Consider the flow of money: Chapter 5, *The Business Context of Communication*.

The days when having cool technology was enough for investors are, sadly, over. You can no longer stick *.com* on the end of a company name or call your product AI and expect investors to throw money at you. You can no longer just throw a new technology idea at the wall and see if it sticks. You can no longer just say "I want the newest, fastest, shiniest computer to work on."

Money is tight, and people who spend it expect their investment to be successful. Engineers need to understand at least the basics of this equation if they expect to convince their leaders or investors to give them money for a project.

Basically, there are two ways to measure the success of an investment: straight return on investment (ROI) and cost avoidance.

Straight return on investment (ROI)

Straight ROI tells you how much you can expect to make on an investment in a specific time frame. This is the measure you use when you're asking for money to build a new product. It is a simple calculation of percent:

```
ROI = (revenue - costs) / cost
```

Personal example

Let's say you want to earn extra money doing freelance web development. Let's also say that you need a new computer to do that. During your first year you buy a new computer for $2,000. You land five projects, each worth about $2000. That means you spent $2,000 and made $10,000.

Using a straight return on investment calculation, your return on investment looks like this:

```
Revenue = $10,000
Costs   =  $2,000 (your COGS, which is the cost of the new computer)
4       = (10,000 - 2,000) / 2,000
```

Your ROI is 400%! You made 4 times what you spent.

Business example

So, let's go back to our birdhouse example from Chapter 5, *The Business Context of Communication*, plug in the numbers, and calculate the ROI for that business:

```
Revenue = $10,000
Costs   =  $7,000 (COGS+GA)
0.43    = (10,000 - 7,000) / 7,000
```

Your ROI on the birdhouses was a healthy 43% in a month.

Remember, though, that's the *total* ROI. Your investors only receive a portion, because you need to keep some of your profit to design the next generation of birdhouses and pay yourself.

ROI also has another dimension: time. Investors want to know how long it will take to pay them back, plus they expect to make a profit for lending you the money.

So, to calculate your investor's ROI, let's say they lend you $100,000 to get started. Let's say you're going to give them 30% of your net profit for the year to repay the loan plus their profit. At the end of the year, how much can they expect to have made?

To figure this out, you have to project from your first month. Let's say your sales per month vary some, and you expect to sell about 100,000 birdhouses during the year. In that case, your projected ROI for the year is as follows:

```
Revenue = $1,000,000 ($10 per birdhouse*100,000 birdhouses)
Costs   =   $260,000 (COGS+GA)
  COGS  =   $200,000 ($2 per birdhouse*100,000 birdhouses)
  GA    =    $60,000 ($5,000*12)
2.85    = (1,000,000 - 260,000) / 260,000
```

You made a very healthy profit margin of 285%, or $740,000.

You promised your investors 30% of that, so if you sell 100,000 birdhouses, they'll get $222,000. Using our ROI calculation, their ROI is as follows:

```
1.22 = (222,000 - 100,000) / 100,000
```

Their ROI is 122%—they got their initial investment of $100,000 back, plus $122,000 in profit.

Obviously, this figure varies if you sell more or fewer birdhouses in the year, but this is the initial number you present, based on your first month.

Cost avoidance

Cost avoidance means what it sounds like: if you invest $X in something, you *save* more than $X over the long term. It does *not* mean cutting costs—it means spending money you were spending anyway, but spending it smarter. This is the measure you use in business when requesting new equipment or software for ongoing, or run-the-business, efforts.

Personal example

If you don't change your oil regularly, your car could run out of oil. If your car runs out of oil, your engine will seize up (not may; it *will* at some point, especially if you have an older car). If your engine seizes up, you will need either a new (or new to you) car or you will need a new engine.

On average in 2024, new engines run somewhere between $5,000 and $10,000 dollars. It might be more if you have a fancy car or less if you decide to buy a used or refurbished engine.

In 2025, the average price of a new car is about $48,000 (depending on what car you get). The average price of a used car in good condition is around $25,000.[5]

Changing the oil in your car costs around $66 each time you do it. For the sake of argument, let's say you change your oil 4 times a year. In a typical year, you'd spend $264 on oil changes.

That means you could buy:

- 19 oil changes for the cost of a new engine (4 years' worth or oil changes)
- 178 oil changes for the cost of a new car (44 years' worth of oil changes)
- 94 oil changes for the cost of a good used car (23 years' worth of oil changes)

You're clearly practicing cost avoidance if you change the oil in your car regularly, because it is much, much cheaper to get your oil changed regularly than to buy new or used cars regularly.

Business example

You're part of an IT group. You manage five servers, and one of them is very old and not capable of running the latest operating system. This means that the server is a little unreliable (and horribly vulnerable to cyber attacks) unless you and your team spend a lot of time applying patches and performing other maintenance work.

---

[5] "The Average Car Price Is Nearing All-Time High" (Luthi 2025)

Unfortunately, that server is the one your source control system lives on, so it must function properly with a solid uptime. Right now, the server is slow and has a lot of downtime while you make repairs.

In a cost avoidance scenario, you would identify all the costs of keeping the old server over a year (it can be any period of time, but let's use a year). Those costs can include:

- Staff time to apply patches. Look at the number of patches applied in the past year and the amount of time spent applying the patches, then multiply by the average hourly cost of your team to get the overall cost.
- Staff time to make repairs. Look at the previous year for a rough estimate of the number and length of the repairs, then multiply by staff costs as noted in the previous bullet.
- The cost of developer downtime—the amount of time the developers spend unable to code because the source control server is down. Look at the past year's server downtime, find out the average hourly cost of the development team, then multiply the server downtime by the number of developers.

Then you identify the costs of a new server. Those costs can include:

- The cost of the hardware.
- The cost of the new operating system.
- The cost of the time to bring the new server up. The total work hours required to bring the new server up multiplied by the average hourly rate of your team.
- The anticipated cost of maintenance—there are always patches and problems; don't try to pretend there won't be. There will just be fewer of them with the new server.

  If you're not sure how to calculate this, take a percentage of the maintenance cost of the old server (pick the smallest reasonable percentage greater than 0).

Then you subtract the cost of the new server from the cost of the old server. That's how much money you save in the first year the new server runs. If you want to get fancy, you can project how long the new server remains new (that is, how long it is before the new server is in the same situation as the old server is now) and project total savings over that time period. However, a year is probably enough.

If you can talk about your idea in a business context, you're ahead of most of the other engineers in the field, and you are more likely to end up with a more successful career. The rest of the corporate team refines the business aspects of your idea, but understanding how to create a business case makes it more likely that they will support your ideas and help you make them real.

Example business case

Let's look at an example email that could provide this information to decision-makers (likely leadership and/or finance). The following example contains all of the elements of a cost-avoidance business case, including a summary table, in a concise email message.

Dear [name of leaders],

I've noticed that we are experiencing productivity issues because our source control server (where we keep all our software intellectual property) is aging.

Our current server is 8 years old, running on Windows Server 2016. IT spends a lot of time maintaining and monitoring this server to make sure it has the most recent updates and is not vulnerable to hacking. Further, Microsoft has announced that it will stop supporting this version of Windows Server in 2027. In addition, we had to upgrade the hard disk capacity twice in the last year alone as the code base expands.

Every time we have to do maintenance, code development stops as the developers must wait until we complete our work to access their code. Performing the maintenance overnight only reduces this problem slightly, as many developers work late or very early. Also, our overseas development team's main working hours occur during our maintenance window.

This aging server is costing us approximately $150K/year.

We propose migrating our code base server to Amazon Web Services (AWS). AWS servers are always running the most recent operating systems, patches, and security upgrades. AWS guarantees at least 99.99% uptime—with no additional IT effort on our part.

You can see the annual reduced costs in the table below.

Item	Current cost (do nothing)	Projected cost (migrate)
Migrate the server contents	$0.00	$5,000.00
Annual AWS subscription costs	$0.00	$14,400.00
Maintain patches and security upgrades	$21,000.00	$0.00
Developer downtime	$129,000.00	$0.00
Total	$150,000.00	$19,400.00
Time to actual reduction in costs	< 1 year	

The table shows that we can reduce our server maintenance costs by 87% and improve developer productivity within less than a year by migrating our source code server to AWS. If you have any questions about my proposal, I am happy to discuss it with you further.

Sincerely,

[your name]

CHAPTER 11
Designing Effective Presentations

You will give many presentations during your career as an engineer. It may be a short presentation to your co-workers on a project you're developing. Or it may be a presentation to an investing group or upper management/leadership asking for funding for a project or business idea (see Chapter 10, *Pitching Ideas*). Regardless of the audience, you need to get up in front of a room full of people and describe and discuss your ideas credibly and effectively.

If you're like most people (74% of the American population, according to research from Statistic Brain Research Institute)[1] you are…less than enthusiastic about public presenting. For most people, public speaking is scarier than flying, financial ruin, sickness, or even death.[2] Public speaking is quite the little anxiety engine.

However, presenting is just another skill. With good guidelines and a little practice, you can get over it—as have many actors, politicians, and other well-known public figures—and deliver quality presentations like a pro. You may never *like* it, but you'll be able to do it. This chapter helps show you how. To dig deeper, see the reference section for this chapter (page 252).

Getting started

The starting point to a good presentation is good material, organized well. That starts with a good slide deck. There's been a backlash to Microsoft PowerPoint and other slide show development tools over the years. We believe that has more to do with bad presentations than the tools themselves. Unfortunately, the tool you use does not prevent you from creating a horribly-designed and badly-delivered presentation.

(i) The guidelines for good slides are also the basic guidelines for good user interface (UI) and user experience (UX) design. Obviously there's much more to a good UI/UX design—it's a whole field of engineering, after all—but these tips can get you started, whether you have to design your own UI or want to have a coherent conversation with the UI/UX team. See Chapter 14, *Constructing Explanations*, for more.

---

[1] "Fear of Public Speaking Statistics" (Statistic Brain Research Institute 2016)

[2] "The neuroscience of stage fright — and how to cope with it" (Dvorsky 2012)

Designing your slides is about a lot more than making them "pretty." Although pretty is one component of an effective presentation (or UI), slide design is *information design*: the act of consciously and deliberately choosing every element on the page (text, images, look, and feel) to create perception and meaning.[3]

If you practice good information design skills, the design of your slides can help your audience:

- Follow your logic to its conclusion: your call to action
- Understand the story your data is telling
- Feel a particular way about your information (the way you want them to feel)
- Examine your ideas critically

(i) If you are working for a company of any size, it's likely that they have a presentation template they want you to use. The material in this chapter applies both to working with a template and to designing your own.

Some important terms

Before we begin discussing the gritty details, let's look at some important terms that pertain to overall design, particularly in short-form documents such as a slide deck. These terms apply to any visual aid you use in a presentation.

Flow: Flow describes how well your audience can follow your slides. Make sure the order of information both in your overall presentation and on each slide is logical and matches the order in which you plan to talk about each concept. Don't make your audience's eyes jump around your slide; lead them from item to item.

Simplicity: Simplicity describes how easily your audience can grasp the main point of your slides. Include only the words you need (more on that later). Make sure any pictures relate to and clearly illustrate your topic. Don't include images *just* to be cute or funny—unless you know your audience very, very well, that never goes as you hope it will.

Clutter: Clutter describes how much excess information your slides contain. In most presentations, you're not restricted as to the number of slides, so if you find yourself with too much detail on a single slide, consider splitting it into two or more slides.

---

[3] For a deeper discussion of information design, see "What Is Information Design?" (Redish 2000).

Reflected light: Reflected light describes how we see ordinary things like a book, a bird, or a tree. The light source is the sun, a light bulb, or light fixture in front of what we're looking at, and the light bounces off the item and into our eyes. Our eyes have evolved over billions of years (all the way back to the first light-sensing cells in the first multicellular ocean organisms), and all else being equal, our eyes and brains are very good at seeing things using reflected light.

Projected light (also called *emitted light*): Projected light describes how we see things on a screen, such as a smartphone, TV, or personal computer. The light source is behind what we are looking at and projected or emitted to our eyes. This is a newer way of taking in light and our eyes have generally not evolved to manage this method as well. That's why you often feel more tired after reading on a screen.

We refer to these terms in various segments of this chapter.

Slides you should always have

Title slide

Every presentation should include a title slide. This is where you introduce the main subject of your presentation and, not incidentally, yourself and your credentials for presenting on the subject. Even if you're presenting within the company you work for, include your name and job title on the title slide. This helps ground your audience and sets a professional tone.

Outline slide

An outline slide helps your audience know what's coming and helps them understand the flow of your presentation. Make the second slide, immediately after the title slide, an outline of your presentation. The outline functions like the table of contents in a book. If you give your audience a copy of your slides before or after the presentation, they can identify what's most important to them. Follow the order of your outline for the rest of the presentation.

Your outline slide should contain only your main points. For example, you can use the titles of each subsequent slide as the contents of your outline slide. If you have several slides on the same topic to continue the flow of your discussion over multiple slides, only include the title in the outline one time. If you follow this best practice and create outline slides using the titles from the slides in the rest of your presentation, we strongly recommend creating the outline slide last, after you know what the titles are.

(i) An outline slide is *not* a slide outline. An outline slide is *not* for you to organize your thoughts. You should have a working slide outline in a *separate document* where you have already organized your thoughts before you ever set fingers to keyboard.

Conclusion

At the end of your presentation, use an effective and strong closing for the next-to-last slide. Your audience is likely to remember your last words best. Use a conclusion slide to:

- summarize the main points of your presentation
- provide a strong call to action
- suggest future avenues of research

Questions?

End your presentation (your last slide) with a simple question slide to:

- invite your audience to ask questions
- provide a visual aid during question period
- avoid ending a presentation abruptly

Handling questions can be tricky, especially if you're not a confident speaker. See the section titled "It's time for questions" in Chapter 12 for information on how to handle questions.

Overall slide deck structure

Whether you're designing your own deck template or using a company or 3rd-party template, the next step is to use your outline (not your outline *slide*, your personal working outline), to build the slide structure. These slides fit in between the outline slide and the conclusion slide.

One of the most painful parts of sitting through a presentation is trying to follow both the slides and the presenter's comments when the slide structure is cluttered or has ineffective flow. Poor slide structure makes your important information hard to consume, and your audience will tune out—they can only take so much cognitive load. Creating an effective, logical, and strong slide structure, on the other hand, increases audience engagement, decreases cognitive load, and increases your confidence as a presenter.

In this segment, we discuss some things to do and to avoid when structuring your slides.

DO

Use 1–2 slides per minute: This gives you easy time markers while you're presenting and can prevent you from rambling because you're nervous or because you're overly enthusiastic about your material. As you do more presentations over time, this metric will change to how you specifically present. For example, Bonni and Sharon know they typically spend 2 minutes per slide and create their slides to that timing.

Write in point form, not complete sentences: Like any other genre, presentation slides have their own grammatical guidelines. The key point of presentation grammar is to only put the most important words on the slide in a bullet-point format. You don't need articles such as *a* or *the*, and in many cases you don't even need complete sentences. Include just enough words to convey the key points of each thought and help keep you on track.

> (i) Using short chunks of grammar rather than full sentences is also a technique that can improve the usability of your product interfaces.

Include no more than 4–5 points per slide: Like using 1–2 slides per minute, this helps keep you on track, time-wise. Try to make sure your points are roughly equal in importance and in the amount of time you need to discuss them appropriately. This helps you avoid spending a large amount of time on one or two points, then looking like you're rushing through the rest.

> (i) Controlling how much information appears in any given interface "page" is also a technique that can improve the usability of your product interfaces.

Use animation to show only one point at a time: This helps your audience concentrate on what you're saying, rather than reading ahead. It also helps you keep your presentation focused. However, there is a big caveat to using animation: it's easy to forget the points you have on the slide. If you display bullet points in this way, make sure you know your material very well, have notes you can glance at (*not* read from), and be prepared to talk around any mistakes.

> (i) You can also use animation when you design a UI. Perhaps a field appears only after a previous field is completed correctly. Perhaps user assistance tips appear on-screen only if the field is not completed correctly. Maybe a button becomes active and changes appearance only after all the fields are complete. Certainly you want to analyze the fields you show on any given screen to ensure they are relevant to the current activity and choose *not* to show fields that aren't. This is called *progressive disclosure*.

Be consistent with animation: Whether you use animation to reveal bullets, to control the appearance of graphics, or to create transitions, be consistent. Choose one animation method and stick with it throughout. And choose the animation carefully. If you are presenting online (and these days you probably are), complex/funky animations may not render or compress well for online streaming.

> (i) Consistency is your friend in slides and in interface design. New information can evoke the fight-or-flight response in some humans—don't make it worse with inconsistent appearance and word choice.

DON'T

Do not use distracting animation or go overboard: Just because your presentation software gives you 30 different animation methods, you are not required to use all of them in a single presentation. The Animation Police are not going to show up and take away the animation options you do not use. Consider selecting the simplest animation (a simple Appear or Fly From Left/Right).

> (i) When the internet was new in the 90's, web pages often included flashing lights, buttons that moved around, or horrific color choices to get attention. Yeah, don't do that.

Don't be wordy: Your audience did not come to the presentation, whether in person or over the internet, to read slides. They came to hear an actual person, in this case you, share your knowledge and *explain* the concepts you're covering. Wordy slides divert attention from you, so the audience spends its time reading instead of listening.

> (i) Your users did not buy your product so that they could spend all their time reading and deciphering words in an interface. Focus on using the fewest words possible—and make sure they are the *right* words. For example, no one wants to "submit" anything. Pick a different, more descriptive word or short phrase for the button action. And then keep that word or phrase exactly the same for that action everywhere in your product.

Text formatting

When you create slides, don't just bang the text on the slide and congratulate yourself on a job well done. The slides are the *user interface* to your presentation—put as much thought (or more) into designing the text of your slides as you do into designing your devices or code.

(i) | Everything we say here about formatting text for presentation slides applies to UI/UX design as well.

Font size

A slide is not a piece of paper (and neither is a computer interface or device control panel). In most cases, your audience cannot pick up your slides and move them closer to or further away from their eyes as they can with a piece of paper. You must make the slides easy to read for as many of your audience members as possible. In general, do not set your text below 18 points.

Font size can also help highlight the information hierarchy of your presentation by, for example, identifying which text is the slide title versus the main point versus any sub-points. We have found that people can easily distinguish size differences of about 3 points (depending on the font), so consider using a pattern such as 18 points for sub-bullets, 21 points for main bullets, and 24 points or larger for slide titles.

Font attributes

Font attributes emphasize certain words or phrases. Typically, you choose among attributes such as bold, italic, underline, or a combination. Each of these attributes signals something different:

- **Bold** is a way of capturing attention. It says "hey, look at me! Look at me first!"
- *Italics* are a way of slowing a reader down. They say "this is important, read me more slowly."
- <u>Underline</u> was originally used as a way in handwritten text to signal the typesetter to set the text in italic. Then it was used in typewritten text where other emphasis methods were unavailable. Currently, it is used to indicate a hyperlink, so we do not recommend using underline as a method of emphasis.
- ***<u>Combining</u>*** methods enhances the emphasis. For example, if you combine bold and italic, it says "look at me first and read me slowly because I am important." If you combine bold and underline, it's saying "I am a very important hyperlink. Maybe click on me and read before you go on." Do not use more than two methods to signal emphasis. The text starts to look like a ransom note and distracts rather than emphasizes.

Avoid using ALL CAPS, in most cases. Most people see all caps as shouting, and you don't want to shout at your audience like a drill instructor. Unless you are a drill instructor, in which case, CARRY ON.

A final word about emphasis: use it sparingly. Always remember, if you emphasize everything, you've emphasized nothing.

Font type

When you're presenting slides, whether in person or online, you want them to be readable at a distance by the majority of your audience. There are two types of fonts: serif and sans serif.

Serif fonts have flat extensions, like flips or bubbles, on key parts of the letter form called *serifs*. Serifs guide the eye across the page, and in typeset text, make it easier for people to read long lines of text. The serifs also help distinguish one letter from another. Times Roman is an example of a serif font. The body text in the print edition of this book is set in a serif font.

Sans-serif fonts lack these extensions; instead they use strong, simple letter shapes to distinguish one letter from another. Arial is an example of a sans-serif font. The headings in the print edition of this book are set in a sans-serif font.

Given these descriptions, you might think that using a serif font in presentation slides would be a fabulous idea. It isn't, though. Remember the discussion of reflected light and projected light from the section titled "Some important terms"? Well, when you're showing slides using a projector, you're using projected light, and that changes the game.

Because the light source for projected light is fairly strong (while it's not as strong as the sun, we still strongly recommend not looking directly into the projector light), the light projected can bend around the letter forms. This bending washes out the serifs, which is a problem because serifs help readers distinguish one letter from another. In most cases, at most sizes, sans-serif fonts are much easier to read at a distance, especially when you display slides using projected light. This also holds true when you use a high-definition TV or are screen sharing.

A last note about fonts and text

Use a standard font like Arial, Helvetica, or Verdana. These fonts are common defaults for a reason: they are highly readable. Further, you may not show your slides using *your* computer, and that other computer may have only the default operating system fonts loaded. If the display computer does not have the fonts you specify, it will substitute another font or display gibberish. Either way, your carefully designed slides will probably turn into an unreadable hot mess

Most importantly, these standard fonts are usually designed for readability, which makes the whole slide deck more accessible, including for attendees or readers with dyslexia or vision issues.

Colors

(i) | If you are using a 3rd-party or company template, consider these color tips within the context of that template. For example, don't introduce new colors; work with the ones designed into the template. Choose graphics that work with the design, rather than fighting the design. Choose table or chart formats that work with the colors designed into the template. The people who created the template spent a lot of time making sure the colors work together—don't muck that up with poor additional color choices.

Color can be an important element of your slide design—if you use it properly. Using color poorly makes your presentation look less professional and, often, more difficult to understand. Colors used poorly make it difficult or even impossible for readers to process information—and poor color choices add to the cognitive load.

(i) | Everything we say about the use of color in presentations applies to UI/UX design as well—maybe even more so.

Color used alone to create informational distinctions excludes anyone with color vision issues. Depending on the population—for example, men are much more likely to have challenges perceiving color than women—from 5% to 8% of people have challenges perceiving color in some part of the spectrum.[4] Low-vision or visually-impaired readers may not see contrast well if your colors are not distinct from each other.

For this reason, we strongly recommend using what are called *redundant signals*. For example, you could highlight an item using color *with* bold or italic text. However, do not go overboard and use All The Colors. You can create massive cognitive overload if you use too many colors or too much emphasis. If you emphasize everything, you've emphasized nothing.

---

[4] "Color Blindness Prevalence" (Mandal 2019)

For example, several years ago, the National Football League (NFL) introduced "color rush" (a special uniform requirement for Thursday Night Football where each team had to wear a solid, NFL-determined color uniform, rather than their regular uniforms). The very first game had the teams in red and green—exactly the same color values of red and green.[5] And the green was almost exactly the same color as the field.[6]

So, to a color-blind person, the game was entirely a gray box, with no distinction between the teams, or even between one of the teams and the field! This was not an effective use of color.

Using color effectively is much more than "making it pretty." Humans respond to color in various ways. Before we explore how to use color, let's examine how color works.

Color is a physical property of items in the world around us. Our world is filled with light, and that light is made up of different wavelengths, which we see as color.

Humans take in color using receptors in their eyes. These receptors process the light (whether reflected or projected) coming into the visual cortex, the area of the brain that interprets sight into signals the rest of our brain can understand. Seeing and responding to color is not merely a decoration; it is a biological process and a way that our brains extract meaning from the world around us.

Responding to color goes beyond merely seeing it, however. Because our brains make use of all the input they receive, we also have emotional responses to color, based on the meaning our culture places on that color. You can trigger these responses with the colors you choose for your presentation—so choose wisely.

(i)   We've noticed over the years that someone, somewhere is teaching that red is a "danger" color. *This is entirely untrue.* If it were true, we would be terrified of strawberries, apples, and so on. Stop signs would paralyze us. What is currently one of the most popular sodas on the market, Coca-Cola, would fill us with unreasoning fear instead of attracting us to a tasty and refreshing beverage. *Red is an "alert" color.* Functionally, it's the same as bold: it says "Hey, look at me!"

---

[5] Color value refers to the saturation and tone of the color—even very different colors of the same value appear as the same gray to certain types of color blindness.

[6] "NFL's Red And Green Uniforms Described As 'Torture' By Colorblind Fans" (Chappell 2015)

This segment is not intended to be a thorough discussion of color and color theory. There are any number of authoritative texts on the subject.[7] Our goals are more practical: let's explore some ways you can use color in slides, visual aids, and user interfaces or control panels.

Contrast

Contrast refers to the difference you can see between colors. Colors that are different from each other are said to have high contrast, for example black and white, blue and yellow, orange and purple. If you've ever used a color wheel,[8] you know that different colors with high contrast are opposite of each other on the color wheel. These colors are also called *complementary colors*.

You can create contrast using *tone*. Tone describes the difference between light and dark versions of a single color. Tones that are very different from each other, for example, light blue vs. dark blue, also create good contrast.

When you create slides, you want them to be readable. In addition to the tips we've already discussed, choosing color combinations that have high contrast against each other increases the readability of your slides.

Figure 11.1 – Low, medium, and high contrast color combinations

Figure 11.1 shows three examples of text against a dark background. You can see that the high-contrast image is easier to read than the low-contrast image. Now stand three to five feet away from your textbook or device. How much harder is it to read the low-contrast image? Remember that your audience will be at least three to five feet away from your slides.

A good way to check the contrast in your images is to follow the accessibility standards set by the World Wide Web Consortium (W3C). Although the full set of standards was created for websites,

---

[7] "The Best Color Theory Books for Foundational Knowledge" (ARTnews 2022)

[8] https://en.wikipedia.org/wiki/Color_wheel

the color contrast standards are good guidelines for print and slide presentations as well. The WebAIM color contrast checker is a web app that makes it easy to check this.[9]

In general, the higher the contrast between background and text, the better. Note that projecting your slides (whether via old-school high-intensity projectors or via screen sharing) can wash out colors, making them lighter on the screen than they appear on your computer. These lighter colors reduce contrast, often to the point where the text is indistinguishable against the background. If you start with very high-contrast colors, you can minimize this effect.

Another way to create contrast is to make sure your slide backgrounds are simple and consistent. *Unless you know exactly what you're doing,* do not over-complicate them with complex images or designs, and make sure the text is clearly readable at a distance.

Structure and emphasis

You can use color to reinforce the logic of your structure; for example, using a light blue title and dark blue text. You can choose this combination because the slide title only needs to be read one time, whereas the rest of the text on the slide carries more meaning.

You can also use color to emphasize a point instead of using a font attribute. A word of warning: if you decide to do this, only do so occasionally, and always select a complementary color with high contrast.

Setting mood

As mentioned previously, humans respond to color. Colors can create mood and trigger emotional responses. Although there can be a wide variety of emotional responses to color (on evolutionary, cultural, and individual bases), here are some general guidelines:

- *Dominant* colors such as red, orange, or yellow usually connote excitement. You can use these (particularly as backgrounds) to rev up your audience, which can make them more excited about your topic. Be careful not to go too far; too much use of dominant colors can be jangly and upsetting.
- *Recessive* colors such as blue, purple, green usually connote calm. You can use these to soothe your audience, which can make them more receptive to your topic if they are doubtful.

---

[9] "Contrast Checker" (WebAIM 2016). In Figure 11.1, only the high-contrast example would pass the W3C's strictest standard (WCAG AAA) when tested using the WebAIM contrast checker.

In general, the brighter and purer the color, the more strongly it makes your audience feel. If you select more muted colors, you can reduce the strength of the reaction, which can be useful, for example, if your presentation delivers unwelcome or upsetting news.

Please note that not all cultures respond to these rules of thumb in the same way. Understand your audience and work within their expectations.

Graphs and charts

To enhance the flow and simplicity of your slides, treat numbers visually, instead of textually. This also reduces the cognitive load.

> (i) You cannot create an effective chart to present unrelated data—and unrelated data doesn't tell a coherent story anyway. Use charts and graphs to compare related data.

For example, instead of writing "four million dollars," write $4M (if inline text) or $4,000,000 (if in equations). This tightens your verbiage and makes numbers more immediately recognizable and scannable. In addition, dollar figures are typically aligned-right so that they can be read more easily. Any time you format an equation visually, align it to the right. Negative numbers (costs) are typically shown in red (see Figure 11.2).

Revenue:	4,000,000.00
- COGS:	750,000.00
= Gross Profit:	3,250,000.00
- G&A:	1,425,000.00
= Net Profit:	1,825,000.00

Figure 11.2 – Tabular representation of numbers

An equation can be considered a type of chart because you are showing data graphically. When presenting other types of data, use graphs and charts rather than tables. Data in graphs is easier to comprehend and retain than raw data is. Trends are easier to visualize in graph form.

> (i) Everything we say here about color applies to UI/UX design as well, particularly when designing reports or data results to be read on-screen.

Here are some common graph types and their uses:

- **Line graphs:** In a line graph, one axis (usually the X axis) is time, and one axis is quantity. The line traces the trend of the quantity over time. For example, Figure 11.3 is a line graph that shows the trend of company sales over time.

Birdhouse Sales This Year

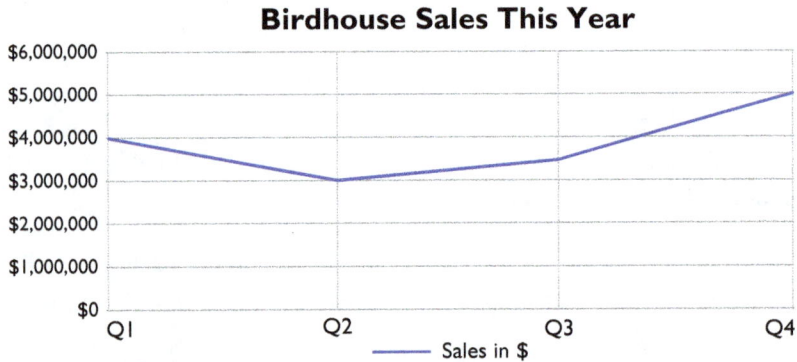

Figure 11.3 – Line graph of birdhouse sales

- **Bar charts:** In a bar chart, each bar represents the quantity of a certain thing, compared on the Y axis to time or some other value. The bars are immediately recognized as a quantity and are easy to compare. For example, you can use a bar chart to show how many units of a product are produced each month (in this case, the number of units produced is more important than the trend over time). You can also use bar charts for easy comparison. For example, Figure 11.4 shows the number of sales for three versions of a birdhouse across four quarters.

Units Sold Per Birdhouse Model #

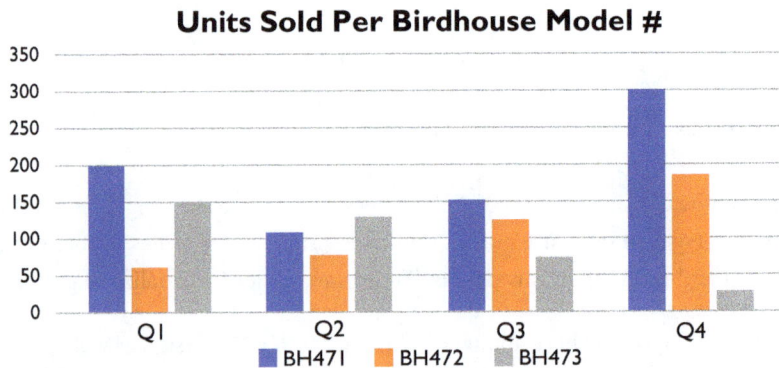

Figure 11.4 – Bar chart of monthly sales

- **Pie chart:** In a pie chart, each slice illustrates the percentage of a whole. Pie charts show the relative importance of a subset to the complete data set, making it very easy to compare subsets to the whole. For example, Figure 11.5 contains a pie chart that shows the percentage of COGS (remember Chapter 5) spent on labor vs. other COGS elements.

Cost of Goods Sold (COGS) Breakdown

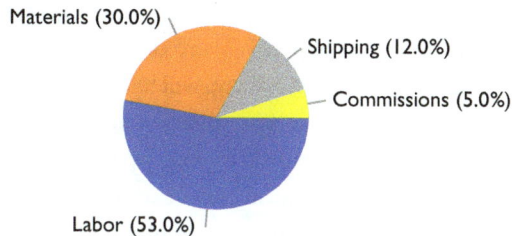

Figure 11.5 – Pie chart

The following considerations can help you create more effective graphs and charts:

Choose the correct graph type

The correct graph type guides your audience to useful conclusions that can drive the decision you want them to make. Poor graph type choice makes your data confusing and obscures the point you're trying to make.[10]

Format the graph well

Remember the concepts of simplicity and flow when designing your graph:

- Use only major grid lines, and as few of them as possible.
- Make sure labels and legends are easily read from a distance.
- Always title your graph or chart.
- Keep the colors and imagery simple.
- Make sure there is high contrast between your data and the slide or graph background.

---

[10] See Samantha Lile's article "44 Types of Graphs Perfect for Every Top Industry" (Lile 2020) for a more detailed discussion and additional examples of chart and graph types. For an even deeper dive, explore thoughts from the master of data visualization, Edward Tufte. *Visual Explanations* (Tufte 1997) and *The Visual Display of Quantitative Information* (Tufte 2001) are a good start.

Spelling and grammar

Proof your slides. Look for spelling errors, grammatical mistakes, repeated words (especially where a line break falls between the repeated words—this is the most difficult type to catch). You're asking your audience to take you seriously as an expert. Every error contained in the comparatively light amount of text in a slide deck undermines that impression.

For example, if you're presenting to a group of venture capitalists, you're probably asking them for a large sum of money. They need to know that you will take care of that money. Simple spelling, grammar, or math mistakes in your presentation make you look sloppy. No one lends money to sloppy engineers. If you can't even get the words in your presentation right with the help of modern spelling and grammar checkers, how on earth can they trust your engineering? As we discuss in Chapter 8, *Ethics in Engineering*, how you do one thing is how you do everything. Get the spelling right.

> We want to be clear that we are talking about more than one or two typos in an entire deck or interface. People can forgive one or two mistakes. They will not forgive a deck or interface that is riddled with errors. And even if they do forgive, numerous errors create a negative impression of your credibility.

The easiest fix for this is to have someone else proofread your slides.

> Proofing is even more important for UI/UX review. If we can't trust you to correct mistakes in your spelling (an easy thing to get right), how can we trust that you fixed the engineering underlying your product (a much more difficult thing to get right)?

CHAPTER 12

Handling Yourself and the Room in Presentations

Engineers do a lot of presentations. They present new product ideas to internal stakeholders, they pitch better solutions to their staff, they pitch entire new technologies to investment people to start a company. One of us worked at a company where the engineers presented what they worked on that quarter to the entire international company. In the business world, you're going to present to a room or screen full of people a lot.

Most people would rather have several dental fillings with no pain relief than present to a room of people. Part of the reason is that few people are trained in how to present to a room of people. Often, you're told you have a presentation next week and figure it out. You do the best you can and, especially if you didn't think it went well, avoid doing it again. But to advance in your career, you must present—it's part of the game.

Being good at presentations can be learned. Every good speaker you see in your life practiced and learned to do it well. Both of us are good public speakers naturally, but we had to learn how to be excellent. You can learn this, too. Use this chapter as a starting place.

Consider taking a stand-up comedy class (Sharon) or an improv class (Bonni) to get comfortable in front of a room full of people. After you've done well making people laugh, you're much more comfortable in the front of the room, and you learn how to handle a room even when you're nervous. And there are tricks you can learn to manage your nerves, even when you think you're going to throw up.

The rest of this chapter includes tips and tricks we've learned to make things go well. Virtually or in person, these tips and tricks make a difference. And remember, you don't have to get international TED-talk good—you just need to be good enough to do a decent job. Knowing this can also reduce your nerves.

A day or two before the presentation

If at all possible, a day or two before your presentation, practice with someone you trust, someone whose opinion will be serious and thoughtful. If you're still in school, don't choose your best friends unless you are sure they won't give you static or mock you. You want someone who can give you actual feedback that you can use. Your buddies may not give you that sort of feedback because they may not take it seriously.

Give your presentation to this test audience as you will give it live. Don't summarize or offer asides (this can mess up your timing), especially if this is an important presentation or if you're especially nervous. Get all the way through the Questions slide before you stop.

After you stop, now is the time for comments. Ask what worked well and what didn't. Let your audience finish their thoughts. Ask follow-up questions, like "So, the money slide was confusing to you. What would help?" Take notes because these comments make your presentation stronger.

It's tempting to ignore the comments because what you may hear, internally, is that you suck, even though *no one* is saying that. If you're not comfortable in front of the room yet, you can hear criticism where no one is intending criticism. You may also be tempted to dismiss the comments because "my audience will know that."

Try very hard to hear what is being said, not the interior voice that humans all have in their heads that says unhelpful things. You're being told what you need to do to make your presentation better. Better in the business world often equals money for your project. It means getting what you want. These comments help you get what you want.

Make changes to your slides and talking material based on the comments you get. Then run through it again on your own until you're comfortable with the information you want to cover. You're more confident and less anxious if you know your material.

Using a script

When you were in fourth grade, your teacher taught you to do presentations from a script. That's because you were a child, and no teacher wants to listen to a group of 9-year-olds just talk about what they're interested in. As you went through school, you continued to use a script because it helped you stay on task.

But now you're in the business world, and you're expected to know the information you are presenting and be able to talk naturally. In the business world, reading a script when you give a presentation makes you look like a child. It damages your credibility and makes you look foolish.

If you're nervous, you may still think using a script is the solution. That way, you don't need to worry about losing track of what you want to say. It's all right there in the script. All you need to do is read it aloud and you're golden.

That's a terrible idea.

Most people are not good readers. Most people are not natural voice actors. You wind up reading in a monotone, stumbling over words, and repeating yourself. If you're really nervous, you may read in that monotone really fast to get it over with quickly. It's death to listen to.

Or you may try to sound excited and wind up using the late night TV sales person voice where everything is exciting and upbeat. You use your voice to be way over the top, thinking that's going to generate excitement for your topic. It's death to listen to.

And no one wants to watch you read something to them. They came for you to tell a story about your topic. You, head down, staring at the paper or device, as you read to them isn't engaging.

OK, you may think, no script. I got it. I'll memorize my script and that way I'll do better.

And that's also a terrible idea.

Memorizing is not going to stop the monotone or late night TV salesperson voice. You may still do either. But now, if you get flustered and lose your place, you have nothing to fall back on. You need to remember where you were when you got lost and mentally get back to that place, while managing being flustered, all in real time, as people are looking at you. That's a considerable ask of yourself. It can trigger anxiety even in people who present well and comfortably.

Don't use a script. Don't memorize. Become familiar enough with your topic that you can talk about it naturally, with perhaps physical note cards with your points in front of you to make sure you covered everything you wanted to cover. Occasionally glancing at your note cards to make sure you're on track is fine. Reading a script badly or reciting a memorized script is death.

Right before the presentation

It's presentation day. You have your presentation set up, you've practiced the material. Now it's time to set up to make sure you have control over the variables you can control.

Regardless of whether you're presenting physically or virtually, get to the room (physical or virtual) 30 minutes early with your presentation device—typically a laptop but tablets and phones are becoming more popular. Bring power cables(s) with you because, no matter how charged your device is, halfway through your presentation, the low-battery light will start flashing, and the battery will go to zero. If you have your power cable with you, this won't happen. The device knows you have the cable and behaves. Or at least that's what seems to happen.

Connect your device to the projector, TV, or meeting software and make sure it all works. If you are physically in a room, there is no standard for how projectors connect to devices. For example, you may need to boot and reboot all the equipment several times to get it all talking together. Even if it all worked the last time you tried it. You want that time to get it all working.

After the tech is working, open your presentation and display it. If you're in a large room, do a quick run to the back and make sure the screen is readable. Because you got there early, you have time to make changes to the slides if the fonts are too small. Remember, a percentage of your audience may have vision issues that make it hard for them to see your presentation clearly if the fonts are too small. The people in control of the money and decisions in your company are likely to be older, with aging eyes, and you want them to see your slides clearly.

If you're demonstrating an application of some sort, open it and get to the place you need to be in the application. If you need another document, open it and get to where you need to be in the document. Minimize the application or document(s). You want these items available and ready to go, but you may not necessarily want them on-screen right away. No one wants to watch you nervously look for an application or document during your presentation. During the presentation itself, when you're standing at the front of the room and your heart is pounding, you will suddenly be unable to find things. Better to have them open and ready for you to use.

Whether you are in a physical or a virtual room, turn off your Slack or Teams notices, and all your messaging/email software. You don't need to share those apps. It can be very distracting to have messages suddenly appear in the middle of your presentation, especially if you have snarky co-workers. Ask us how we know.

Bring a notebook and a pen or something to write on. Of course, you have a physical notebook that you use to take notes to yourself in the workplace, keep lists, and track action items. Bring that with you and a pen or pencil. You may need to use it at the end of your presentation. More on that later.

> (i) In the workplace, you may want to take notes only on devices, like your tablet or phone. There is nothing wrong with that. However, during a presentation, seriously consider using a physical notebook in the moment and then typing important notes into your application of choice later. It's often easier to quickly write notes in a physical notebook than to get your phone out, open the application, find the right area, then type on the small keyboard. When you're talking to someone or in the front of the room, making quick notes in a notebook is much faster and easier.

The presentation

It's showtime. It's time to start your presentation.

You may have heard that you should start a presentation or talk with a joke to warm up your audience. Unless and until you're a very confident presenter, please don't do that. One of us (Sharon) has done stand-up comedy and still can't tell a joke. She can tell a funny story, but remembering all the parts of a joke is too much. She's the annoying person who gets halfway through a joke and then says "Wait, there was a penguin, let me start over." Or she gets to the end of the joke and says, "and then the penguin said…I forgot what the penguin said. Wait. Hang on." It's a mess. She's so bad that the telling of the joke becomes a performance in itself.

Fortunately, she's very comfortable blowing a joke in the front of the room and can easily recover from this going terribly wrong. She's done stand-up comedy and has died on stage. If you're nervous or uncertain, your opening joke going wrong can so unsettle you that recovering is too much. Don't set yourself up for a possible failure, especially if you're nervous.

OK, you may be thinking, but I'm hilarious and I can tell a joke. I want to start with a joke, despite the warnings. Please, don't do it. You may be able to tell a joke well, remembering all the parts all in the right order. But what are you going to do if it's the wrong joke for that audience? What are you going to do if the room is silent when you're done? If you're at all nervous, you may decide to explain the joke, which always makes a joke funnier (not). Just avoid jokes until you're a very experienced speaker who can read the room.

Manage yourself while you talk

Introduce yourself by name, unless you know everyone in the room knows who you are. Thank them for coming and start talking. If you're nervous, say "I have a lot of information to get through, so please hold your questions to the end." This sounds like you're concerned with your audience and want to make sure they get all the information you have to tell them. But what you have done is asked them to not interrupt you and throw you off track until you're done. This can help you because random questions can throw you off your prepared stream of talk. If you're nervous, it can be very difficult to get yourself back on track.

Use your body

Obviously, if you're presenting virtually, this section is less important. But some of these ideas can be adapted to virtual meetings with or without a camera.

While you talk, use your body well. Face your audience and look at them, even if you're virtual. If you're nervous, this may be hard. You may want to turn your body away from the audience until you have almost turned your back to them. Think about what it would be like to talk with someone who turns away from you the entire time. That would be weird, so don't do that.

Look at your audience (or camera) while you're talking to them. But, you may be thinking, I don't want to look at them looking at me looking at them. It makes me uncomfortable. If you're physically in a room with the audience, you can use the Z sweep to look at them without actually looking at them. Start in either back corner and slowly look at foreheads across the back row. When you hit the end of the row, move your forehead viewing diagonally to the front of the corner of the room. Finish by looking at foreheads across the front row. Humans can't tell you didn't look in their eyes because you're looking just an inch above their eyes. They feel looked at, and you avoid the scary feeling of looking in people's eyes. Do a Z sweep every few minutes.

Make sure you look at all sides of the room, especially if it's large. Keep in mind you may tend to talk to the right- or left-hand side of the room depending on whether you're right or left handed—we are both right-handed and often want to talk to our right side of the room. If an area of the room is smiling and nodding, you may find yourself talking just to them. Watch that and make sure you include the entire room. Think about how it looks if you just talk to one or two people in a crowd of twenty. For an hour. That would be weird, so don't do that.

Stand up straight (or sit up straight in a virtual meeting). If you're uncomfortable, you might want to slouch, thinking that makes you look smaller. One of us is just over 6-feet tall. Trust her when she tells you slouching doesn't make you look smaller; it makes you look as though you recently had a back injury, and now that's all anyone is thinking about as you talk.

If you naturally move when you stand, feel free to move if you're presenting in person. If you decide to move around while you talk, try to stop moving when you're making a point. It focuses the attention in the room on you. If you naturally use your hands when you talk, feel free to use your hands. Do the things that are natural for you when you talk because trying to artificially use your hands or body looks odd. It's not how you inhabit your body normally and it shows.

If you don't want to use your hands, you can put them behind you in "parade rest." Grasp one wrist with the other hand and loosely hold them behind you. If you want to put your hands in your pockets, leave your thumbs out. Don't play with anything in your pockets because your audience can see you. If you're nervous and you put your fists in your pockets, you may clench them tightly, making your hands swell. At some point you need to remove your hands, but they may not come out of your pockets easily. You can look like you have a fight going on in your pants. That would be weird, so don't do that.

Use your voice
Speak up clearly as you talk. The back of the room needs to hear you. If you're not confident you can project your voice for the entire presentation, use a microphone. If you do use one, practice first. Hearing your own voice come back to you from speakers can be unsettling. You can get used to it pretty quickly, but it can throw you initially. If you're naturally a soft speaker, seriously consider a microphone. You want your brilliant ideas to be heard.

If you're presenting virtually, don't whisper; speak at a normal volume. Yes, people can adjust their speakers, but you still need to speak at a normal volume.

Avoid *upspeak*. Upspeak is when you raise your voice at the end of every phrase or sentence, as though you're asking a question. Worst case is when you raise your voice every three to five words in a sentence with a pause. You sound as though you're confirming? everything? you're saying? with your audience? as though they know? and you don't? *You're* the expert in what you're presenting, so you don't need to confirm every five words with your audience. That would be weird, so don't do that.

Watch your patterns of speech. It's very common for people who are nervous to have verbal tics, such as saying "like" all the time. We both say "So" and "right" far too much and both of us work to remove those tics from our speech. Saying "like" every three to five words is very distracting to your audience. And you sound as though you are unsure about your topic.

Don't read your slides to your audience—you're telling a story from your slides. If you have to read your slides to your audience, you're not prepared to tell the story you're here to tell. Your slides are the main points you want your audience to remember. See Chapter 11, *Designing Effective Presentations*, for how to put together good slides.

If you have more than one presenter, don't leave the physical or virtual stage empty, even for a moment. When the center of the room is empty, you lose the attention of your audience, and getting it back can be hard work. Don't make that sort of work for yourself. If you're presenting virtually, make your handovers smooth, perhaps by having one person run slides for everyone. In the few moments you use to change presenters, people check their email and check out.

It's time for questions

Your last slide should say **Questions?** so people know they can ask questions now. Turn to the room or the camera and ask if there are any questions while you quietly check your time. In your head, slowly count to five.

If you're nervous, questions can be the most scary part because you just handed control to the audience, and you have no idea what may happen. But there are ways to cope.

First of all, you don't have to have all the answers to everything. No one expects you to have the answer to every possible question. There is nothing wrong with saying "I have no idea, I've not thought about that. Let me do some research/look at that/other thing and get back to you." This is where your physical notebook is your friend because you can open the notebook and make a note to follow up with that person. And of course, you will follow up.

If you're presenting virtually, you may get questions in chat. That takes longer for people to type. Waiting five seconds may not be enough time. One of the ways we manage this in teaching a large virtual room of people is we ask people to use the Raise your Hand feature and then type their question. That lets us know a question is incoming, so we can be patient while the person types it. Consider a similar method in your virtual meetings to manage the lag time. And of course, read the question out loud for the attendees before you answer it.

Call on the highest-ranking people in the company in the room first, if you know who they are. Very often, they have one question and need to get to the next meeting. Show them you respect their time and take their questions first.

Etiquette and manners

Regardless of the person's rank in the company, allow the person to ask the entire question. It's very annoying to get five words out and be interrupted by a presenter who barrels forth with an answer to what they think your question is. Both of us are so tired of this behavior that when it happens to us, we wait politely while the person drives all over answering and then say: "That's nice. What I was asking though is…." We know that's not very polite, but we've reached a point where we just can't do it anymore. Respect the person who is asking the question enough to let them ask their question.

If you're nervous or don't think on your feet well, develop the habit of pausing and then restating the question. That gives you a few seconds to think about how you might want to answer. It also helps you if you get a confusing question. Restating a confusing question to reflect your understanding can help the questioner clarify the question. It can also help you direct the question to what you wanted to be asked, which can be very helpful sometimes.

Restating the question is *essential* if you're in a large room and you're the only one with a microphone. If you're presenting virtually, reading chat questions aloud and then answering them is also helpful to the attendees.

With an eye on time, continue to ask if there are questions. In your head, count slowly to five. If there are no more questions at that point, thank your audience and remind them that they can reach you at the email address on your last slide if they have questions later.

Politics in presentations

During the question and answer part of your presentation, it's possible to have politics come up. Let's look at several scenarios that can happen and some ideas for managing them.

I suggested this x years ago and was told no. Notice this is not a question. That's because this is not a request for more information—it's to let you and the room know that person had the same idea and was shot down. If the room seemed interested in your idea, this can also be a way to let you know you're not so smart. Publicly. The only way we've found to deal with this is to act excited and ask the person if you can meet offline to share ideas. Write down their information in your notebook and follow up. Sometimes they have ideas that can contribute to your idea.

Sometimes you can learn why the company declined the project at that time and use that to push your idea now. Technology changes and directions of companies change.

Discussion of other, often unrelated, topics occurs around you. This can happen when someone asks a question that sparks an idea from someone else, and then they're deep in discussion, totally ignoring you. This is especially difficult if these people outrank you in the company, because interrupting them can be considered rude. If you're uncomfortable interrupting, try to catch the eye of someone of a similar rank if you can, and do the eye thing. If this is a decent work culture, they will step in and redirect for you. If this is not a good work culture, that person may look away and not help. If that's the case or if you can't catch the eye of anyone suitable, you need to step in and suggest the discussion be taken offline so you can answer the rest of the questions.

No questions at all. This can be very disconcerting. You ask if there are questions and everyone looks at you quietly. Don't panic—if you're near lunch or right after lunch, people may not have their thinking caps on. If it's late in the day, people may be fried. If you're near a release, people may have their heads in another place. And it could be that there were just not any questions at this time. It happens. Be polite, slowly count to five in your head, and then thank them for coming.

Finishing up

Now that your presentation is over and you thanked your audience, start packing yourself up, if you're physically in a room, or shutting things down if this is virtual. People may want to come ask you questions and chat while you're packing up. This is informal chat, although you may be asked quite serious questions about your topic. Continue to pack up as you chat, but listen closely.

If the presentation is virtual, be the last person to leave the meeting so anyone who types a question doesn't lose you. If you know the meeting room has people waiting to use it, as happens in many workplaces, ask the person if you can continue discussion in the hallway, so the next group can use the room. Finally, if people have left things on the tables, clean up. It's just good manners to leave the room clean for the next meeting, even if it wasn't when you walked in.

CHAPTER 13

Cognitive Science

Because you're building products for humans, it's a good idea for you to understand how humans structure the world in their heads. Even if you're building hardware that will only be used by other engineers as components in their products, those engineers are humans who need to know how to use your hardware.

At some point, humans need to decide if your product is suitable for their purpose and then get it to work to serve that purpose. Even if your product becomes part of another product and isn't directly used by people, it still must work with that product. Middleware is a software example of this sort of product, as are raw materials and physical parts for a hardware product.

If you're building products that customers directly use, such as an app or a Software as a Service (SaaS) product, understanding how humans construct and think about the world is crucial.

This chapter is not a deep dive into how humans construct the world in their heads. It is an overview of what we have found to be useful to know as you design products and communicate with people. If you want a deep dive, we strongly recommend you get a textbook or take a course about cognitive psychology. The reference section for this chapter (page 252) has some good books listed as well.

About humans

Humans, in all times and all places, have to deal with the physical world. Humans are not alone in this—all creatures need to do this. Salamanders must figure out the world well enough to negotiate their physical environment, not get eaten, find food and moisture, mate, and lay eggs. They do not need a robust understanding of their world. Luckily, evolution is conservative, and they generally know enough to survive, at least in the aggregate.

Salamanders live about 20 years—if they don't get eaten or develop a disease—which is much shorter than most humans live. They don't travel as much as humans do, which was the case even before industrialization. While salamanders are cool, they don't need to be more than they are.

Humans are more complex. Much more complex.

Humans need systems of information that provide all the things salamanders need, but in a longer time span. Humans have language, which allows us to encode and share information about things and events other humans have not seen or experienced. Humans also have culture, which encodes a lot of other things that help us survive, such as what is good to eat, how to dress, who to marry, why things happen, how to earn a living, and so on. Our cultures also tell us important information, like what's real and why.

Remember when we talked about red as a *danger* color in Chapter 11, *Designing Effective Presentations*? Red as a danger color is cultural—in other words, your culture may say red is dangerous. That's culture, not biology. Biologically, red is an *alert* color.

Variety in human behaviors tells us this is cultural, not biology. For example, biology means humans must consume calories; culture tells us not to eat insects. Or it does tell us to eat insects, but avoid other things that are, in fact, edible. The attributes of what's disgusting to eat is generally driven by culture, otherwise all humans would think the same foods are disgusting, which we certainly do not.

The physical world and biology

Culture tells its members what's real, how to make a living, what they should or shouldn't eat, what music they like, how to dress, and many other things. But all this has to be grounded in the physical world, working with our biology.

For example, culture can tell members that as part of the culture's rites into adulthood, all teenagers must climb that cliff and then fly down to join the adults at the base of the cliff. This is not a great strategy if they want to have the culture last for another generation. No matter how much humans want to think they can fly, they cannot fly without the aid of a great deal of technology, even if their culture tells them they can.

As humans, we must interact with the physical world we live in, and we must manage our biology in the physical world. For example, if you live in the arctic, you must wear clothing to keep warm enough to not die or lose too many body parts. You would be used to a certain level of cold, but you still need appropriate coverings in the winter. Without it, your biology is endangered and, in the worst situation, you die.

Which brings us to how humans biologically perceive and interact with the physical world.

Our vision

People use their senses to identify their environment and to interact with it. However, humans are locked into the limitations of human perception systems. For example, without additional technology, humans cannot see ultraviolet. Bees see ultraviolet, but they cannot see red. Humans see red very well because of the way the rods and the cones in their eyes are constructed.

Humans see red so well that if a language has only three color words, the colors are light, dark, and red.[1] The range is what one would expect, based on the way all humans see color. If a language has only four words for color, they are light, dark, red, and either yellow or green. If a language has five words, it includes both yellow and green. And if it has six words, it adds blue.

The point here is if these are the first colors humans identify by name in language, it's possibly because the rods and cones in human eyes see those colors really well. This is helpful to know when you're designing products. You want to use these colors for things that are important because people see and identify these colors so well.

You also want to make sure you communicate appropriately with color. For example, red is not a danger color in all cultures or subcultures. In some Asian cultures, red is the color of joy/luck and can be the correct color for a wedding dress. In the subculture of accounting, red means deficit. In some cultures, yellow is the color of spring and joy while in others it's a sign of cowardice.

When you're designing products, be aware of the colors humans see very well and the cultural considerations of those colors. A Google search for international color meanings can help you avoid the cultural faux pas of communicating the wrong message with your colors.

The world happens in human brains

Of course, humans don't see anything with their eyes; eyes are a window for light that gets sent to the brain, where the brain decodes and assigns meaning and relevance. Human eyes are not the only way brains get signals, but it's the way we are going to focus on in this chapter.

Preattentive process

The *preattentive* process refers to automatic behavior and perception that don't require thought. This process doesn't use higher cognitive functions to accomplish a task or interpret input. It's

---

[1] *Basic Color Terms* (Berlin 1969). Examples include several languages in the highlands of New Guinea.

the automatic pilot that happens, for example, when you get in your car and find yourself heading to work when you meant to go to your aunt's house.

However, preattentive processes are deeper than that. They can also be thought of as knowing where and how your body exists in time and space. For example, you know that when you pick up a cup, the cup isn't your hand. You know they are different. But you didn't always know that— for babies, learning that the cup is not their hand is an attentive process.

When and whether something becomes preattentive in our brains is *not* a conscious choice; it just happens. As designers, we can *influence* that with design choices that leverage the other cognition concepts we discuss in the rest of this chapter and in Chapter 14, *Constructing Explanations*. But no one is making a conscious choice for something to be preattentive.

Attentive process

The *attentive* process is, as the name suggests, more front of mind. The task or perception requires cognitive effort, such as learning, memory, and understanding. The attentive process is slow and includes a great deal of cognitive effort. Learning to read is a good example of an attentive process. Watch a small child read aloud, and you can see how slow this process is and the effort it requires.

But children are not the only ones in the attentive process. This process occurs in humans all the time as they try new things in the world, such as using a new recipe that requires unfamiliar techniques. As you become comfortable with the recipe, you don't need to concentrate as hard the next time you make it.

The attentive process is a cognitively expensive place to work in. It's hard to stay in the attentive process because it requires so much of your available system resources. Keeping people in this mode requires your readers to stay in a difficult place and mentally exhausts them, reducing their ability to comprehend and act.

Combining preattentive with attentive

These processes can be combined because human brains are complex. Activities that start out as attentive can, with repetition, become preattentive. For example, driving a car starts as a totally attentive task with many layers of perceptions and a very high cognitive load. The first few times you drove a car were exhausting, and you may have actually taken a nap when you were done. Your brain needed to stay in attentive mode for a long time, and that's exhausting.

But now, you can drive home and realize when you pull into your parking space that you have no idea how you got home. You have no memory of the drive at all. But the car seems undamaged, and no police officers are behind you, so it must have worked out okay.

However, things that are preattentive but now need to be attentive can get very messy. For example, Sharon was in New Zealand for about a month for work. She had a rental car and drove five days a week from the hotel to the work site and then back to the hotel at the end of the day.

Her drive was cognitively exhausting because New Zealand drives on the opposite side of the road from the United States. Therefore, she had to attend to everything. When possible, she followed the car in front of her, which helped. But every day, after driving the 20 miles safely back, she made the last turn onto the small street where the hotel was located. And every day, she found herself driving at cars in her lane. She quickly corrected, but many people went home each night with a story about another tourist driving badly.

So, why was the last quarter mile of that drive always a mistake? Because driving is normally a preattentive activity, but it had become an attentive activity because Sharon was driving on the other side of the road. The cognitive load was high and continuous. At the first chance, her brain dropped back into preattentive because it was tired—and it was tired at the end of the drive. It's cognitively very hard to be in attentive mode when the task or perception has been preattentive.

So why do you care about this when you're designing products? Because your users are counting on you to make products that don't keep them in a high cognitive load for long. How can you immediately reduce cognitive load?

Standards are preattentive

Seriously, standards.

If you're making software or apps, the operating system you're coding for has standards for how your product should look and behave. Use them. Your audience is counting on these standards to give them a basic grasp of how to interact with your product.

You may not agree with these standards. You may think these standards are stupid and you can do better. You may even know of usability research that shows something in the standard is a bad idea. We get it. We feel your pain.

But your audience expects your software or app to look and behave (to communicate, if you will) just like all the other ones they use. If you follow those standards, you immediately reduce the cognitive load. You provide the audience with the comfort and assurance that they have some familiarity with what you're doing. They feel like they might be able to do this.

Additionally, especially if you do not have access to a usability group or research, the companies that created the standards probably did have that access. They often did the research to come up with the standards. They studied users and made decisions based on how people used their products. Leverage that knowledge in your products.

Standards are not going to solve all your problems—your user may still struggle with your product. It happens. But by leveraging standards, you solve your users' the first challenge. You allow them to use their preattentive state to use your product, leaving more of their brain available to be attentive to the things that are unique to your product.

Other processes

While there are many other processes that go on in our brains, we focus on two more here.

Cocktail party effect

The cocktail party effect is a way for human brains to filter out what they don't want to attend to and focus on what they do want to attend to, despite the environment. It's named for the effect of being at a loud party and still being able to have a conversation with someone despite loud music and lots of people laughing and making conversation around you.

Humans do this all the time. You hear your name in a loud room, despite not hearing any one specific conversation. The cocktail party effect protects you from the onrush of data coming at you in a noisy environment and allows you to focus. Your brain takes multiple streams of information and picks one to attend to. It's a really neat trick that human brains do, and it's a trick your brain just decides to do.

Why do you care? You want your users to focus on what you need them to focus on and pick out what you want them to pick out. You need to think about the environment they are working in— the scenarios in Chapter 15, *Personas and Scenarios*, help a lot here—and design the product to accommodate that environment. You need to get and keep their attention, especially with error states, when things have gone wrong and you really need them to attend to what's gone wrong.

Sensory adaptation

Sensory adaptation is the process where humans cease to attend to a stimulus after long exposure. It's basically the opposite of the cocktail party effect. People no longer hear or see the stimulus. Their brain decides that the stimulus isn't important and filters it out.

For example, the Southern California city one of the authors lives in has a deep history of citrus production. 100 years ago, railroads were built to get that citrus to the rest of the country. That means we have many railroad tracks. Today, we have far less citrus, but those tracks still ship lots of goods from the ports of Los Angeles and Long Beach to the rest of the country.

In most of the central (older) part of the city, she can hear the rumble and horns from all those trains. Except she doesn't really hear them anymore because her brain has adapted to that sensory input. She only hears the trains when something unusual occurs.

Like the cocktail party effect, sensory adaptation is not something humans have a choice about— it's just what brains do. If you decide to have all alerts in your product flash red, it's likely your users will adapt to the flashing red, and after a while, their brains will ignore alerts. This is not good if the alert means the reactor water is low. You need people to attend to that alert.

You can solve this by using a unique delivery method for the most important alerts. For example, fire alarms are very loud and hard to ignore; further, they are usually coupled with redundant signals such as flashing lights. Nothing else sounds like that in people's daily lives, and this is by design. The consequences of not hearing or seeing the fire alarm are significant. For critical errors where lives may be in danger, make the alert so different that it's impossible to adapt to it.

Learning theory

Related to the previous topic is learning theory—how humans experience and process the world. We are concerned about:

- Experience
- Schemas
- Habits
- Reinforcement
- Interference
- Cognitive load

These concepts don't stand alone and are tightly related to each other, so the following sections are full of cross references. As with many things about the human brain, this is a little messy.

Experience

Learning is the result of experience. We don't mean sitting in a classroom, listening to a lecturer drone away. We mean the learning that happens as you go through your day, interacting with the environment, people, and everything you do.

As you go through life, you add to existing learning and learn new things. As you do this, you gain experience, from which you draw as you interact with the environment and people around you. The experiences you draw from are called *schemas*, which we discuss in the next section.

Human brains are plastic and learn all the time. Experiences drive what we do. They are both common and unique. They are also partly culturally determined, or at least structured. For example, the way you greet your grandmother in India may be very different from how you greet her in another culture. You learned how to greet her through experience and learning.

Schemas

Schemas are the result of all that learning and experience. They are the stories humans create that explain the world. People develop mental models that explain the world and structure their interactions with the world. For example, how you greet your grandmother is determined by the schemas you have about at least the following:

- Greeting people
- Greeting family members
- Greeting older people
- Greeting older family members
- Politeness
- Your age

The intersections of all these schemas determine how you behave when you greet your grandmother. It's unlikely you punch your grandmother in the arm when you greet her, certainly after you get to school age. It's also unlikely you kiss or hug her the way you would greet a romantic partner. It would be weird to do that. These rules you know without thinking are *schemas*.

But these schemas shift and change as you gain experience. The way you greet people changes as you get older, as does the way people greet you. At some point, if you're lucky, you become the grandparent being greeted. And you know exactly how people should greet you.

Your users have schemas, too, especially when it comes to approaching your products. You can make some guesses about these schemas, and personas and scenarios help (see Chapter 15, *Personas and Scenarios*). But your job is to design products that fit your users' schemas well enough that they can use your product to do what they need to do and also extend and change their schemas.

Disruptive technology changes the schemas entirely, and this is not always good. Think about the technology adoption curve and early adopters—they may have more extensive schemas around this product space or technology. They are also more open to spending serious time learning your product, possibly because they already have a schema that says "smart people spend serious time learning products."

But the users in the majority of that curve may not be willing to put in the time and effort to massively change their schemas. They certainly will struggle with your product because it's a significant cognitive load to massively change or generate new schemas. Humans can do it, but it requires serious cognitive effort. People need to be motivated to put that effort in, and your game-changing product may not be that motivating for them. Or, they may have a schema that says "only dumb people have to struggle with products."

Standards use existing schemas to help your users learn the new schemas for your product. This is important. It reduces the cognitive effort and allows existing schemas to be used and extended, making it easier to use your product. For example, although her mobile ecosystem is Apple-based, Sharon's personal computer is a PC. She's been a PC user for decades. A recent job standardized on Apple, meaning she had to shift her computing schema. The first month was spent changing her computing schemas, and it was cognitively very difficult.

You can also use schemas when you communicate by using examples and metaphors. Examples and metaphors that resonate with your audience resonate because you tapped into their schemas. Examples and metaphors that don't work for your audience missed their schemas.

By knowing the personas and scenarios for your product, you have a better chance of hitting the schemas. By knowing your audience, you can use examples and metaphors that tap into their existing schemas. Your communication is more likely to work. For more details about examples and metaphors, see Appendix A, *Metaphors*.

Habits

Habits are those conditioned responses humans don't think about anymore. You went through the effort to learn whatever the behavior is, and now you don't think about it to perform it. You went through attentive effort, built a new schema, and instantiated the behavior well enough that you don't need to think about it anymore. You can just do the behavior. Habits are learning that has become preattentive through experience.

For example, you don't think about how to use the most common features in your phone. You just know where everything is and how to get to it. If you need to use someone else's phone for some reason, you're suddenly at a loss as to where anything is. You drop from preattentive into attentive, and your schema is of little help. Your habit doesn't work.

Your users come to products with habits in place, too. If your product has competitors, and most products do, your customers may have habits from other products. You need to support those habits as much as you can to help your users feel more comfortable with your product.

The more comfortable your users are, the more usable they will find your product. Leveraging your users' schemas and habits makes them more comfortable and reduces their cognitive load. One important way to do this is to use the standards defined for your operating system.

Interference

Previous learning can interfere with new learning and building new schemas. Your habits and existing schemas get in the way. They create additional cognitive load. For example, Sharon uses Apple products for her mobile ecosystem. On a whim, she bought an Android tablet, and it was a crashing disaster. It's similar but different enough that many of her existing schemas and habits made learning the Android tablet nearly impossible. She finally gave it away because it was just so much harder to use (for Sharon) than her Apple tablets.

Cognitive load

We've mentioned cognitive load multiple times throughout this book so far. Cognitive load is all the stuff going on in a person's head at any given time. You have no control over the cognitive load your customers bring to your products. They could be having a terrible day, dealing with a sick baby, worrying about an older relative, tired beyond words of pandemics, desperate to be outside on a beautiful day, or dealing with all the other things they may have going on.

But you, as a product designer, have a *lot* of control over the cognitive load you add to your users. You can design a product that is easier to use. You can write clear instructions, create a simple user interface, and provide a clear path for users to complete tasks with your product. Using the things you learned in this chapter, you can reduce the cognitive load you add to the user.

Humans require explanations

Humans are hard-wired for explanations, for a story that explains what's going on. This seems to be how human brains work, and people have no choice about telling these stories. Humans need to understand the world around them. It's a good trait, as it probably kept individuals from being eaten when they were small hominids wandering the savanna.

People construct long, involved stories to explain things in the world.

Your job is to create products that allow your users to tell a story that's close enough to how the product technically works that they can still use the product to meet their needs. Everything in this chapter and the next helps you understand humans well enough to do that.

CHAPTER 14

Constructing Explanations

One of the key purposes of business and technical communication is to explain technical concepts to those who may not have an engineering background. Whether your readers are internal (see Chapter 9, *Flow of a Project in a Company*) or external (end users), they are looking for clear and understandable explanations. Your engineering ideas are only as usable as your communication makes them.

Explanations and our brains

Humans do not like the unexplained. People's brains seem to be wired to construct explanations, and they are deeply unsatisfied if they do not get these explanations. This seems to be biological in nature, rather than cultural. For example, every culture has an origin story that explains where humans, and the world, come from.

Interestingly, in the absence of a clear and simple explanation and left to our own devices, we prefer complex explanations over simpler ones.[1] The less informed one is about a subject, the more complex the subject appears and, thus, the more complicated the uninformed seem to need the explanation to be. And we project that less-informed perspective onto explanations. We feel that if the explanation is easy, we may be stupid for not already seeing it, and thus we often don't trust the simpler explanation.

But needing to know "why" is a fundamental part of who we are as humans. Neil deGrasse Tyson often explains that humans have a fundamental need to "know why" because it's deeply ingrained in our evolutionary development. This need allows us to understand and navigate our environment, predict outcomes, and ultimately, survive. Essentially, our curiosity about the world around us is a key factor in our intelligence and ability to adapt.[2]

---

[1] "Explanatory preferences for complexity matching" (Lim 2020)

[2] "Neil deGrasse Tyson explains how aliens could be so much smarter than us" (Tyson 2017) and "Neil deGrasse Tyson Explains Why Some Info Is Need to Know" (Tyson 2020)

This is so pervasive that it appears to form part of the core of what makes humans different from other animals. And, in fact, coming up with theories about why and how humans are different from other animals is a very human thing to do, with a long history.

However, due to extensive research in the last few decades, many of those previous theories about what makes us special have been abandoned. Research shows that other animals make and use tools,[3] so "Man the Toolmaker" won't work. "Humans are the only animals that recognize themselves as individuals" has been disproven using the mirror test. A variety of studies on songbird and dolphin call analysis that seems to indicate some calls include unique identifiers that function as individual names.[4] Even "Man the Complex Problem-Solver" is taking hits from recent research on corvids (ravens, crows, and jays) and psittacids (parrots), which shows their ability to recognize and solve complex, multi-stage problems.[5] Research on apes, ceteceans (dolphins, whales), and ravens show that these creatures are capable of creating economic systems—and an economy is about as complex a daily-life system as it gets.[6]

So researchers are currently left with "humans really like explanations" as one of the remaining factors that makes humans unique, and humans, as Neil DeGrasse Tyson points out in the remake of *Cosmos*, are pretty desperate to be unique. It will be interesting to see what the next few decades of research show.

Fortunately, this means that humans are very responsive to being given an explanation of the world around them and the items in it. This means you, as a communicator, can and should control the explanation of your ideas, making them simple enough to understand and complex enough to be believable.

The rest of this chapter discusses two ways you can construct those explanations in written documents such as user assistance and specifications (yes, your fellow engineers are humans, too) as well as in user interface and experience design.

---

[3] "Does Tool Use Define Humanity?" (National Geographic 2022)

[4] "Which Animals Recognize Themselves In Mirrors?" (Seeker 2015)

[5] "Are Crows the Ultimate Problem Solvers?" (BBC Earth 2014)

[6] "Dolphin Capital Theory" (Tabarrok 2018) and "Apes can make rational, economic decisions as humans do, research finds" (Firtina 2022)

Design and perception: perceptual explanations

As we discuss in the previous chapter, human beings (any living creature, really) experience the world through their senses. The external input we get from our senses drives the internal processing that forms our perceptions. Strong information design, whether it's for slides, documents, or UI/UX, depends on leveraging human perception effectively. A large part of strong information design rests on understanding how to construct explanations that leverage human perception.

Before delving into how to create perceptual explanations, we need to define a term. For the remainder of this book, *page* means "any visual collection of items that can be seen at a glance." Under this definition, any book page (like this one) is a page. A web page, as the name implies, is a page. Any given screen in a software interface is a page. Any device control panel (whether analog, with physical buttons, or digital, with a touchscreen) is a page. All pages can (and should!) be designed.

All else being equal, humans are strongly visual animals, especially when they are interacting with technology. Human hearing is so-so among the animal kingdom, human sense of smell is a disgrace. Most humans don't go around licking things or putting them in our mouths to gain info, at least not after we turn three or so. While people do indeed continue to touch things, touch is not most humans' primary sense. For most people, eyes are the primary connection to the world, although, as with anything else, humans are nuanced (see the section titled "What are preferred input modes?" in Chapter 15).

> Our human visual focus does *not* mean that "sticking a picture in" makes for good information design. Humans use *everything* on a page—whether that page is in a slide deck, a document, or in a physical interface—to perceive meaning. So as you develop your communication, you need to remember that how it looks might just be as important as what it says. In Chapter 11 we discussed what makes a good presentation slide. Most of these same principles apply to the pages you create, and this section explains why they work.

The phrase *intuitive design* gets thrown around a lot in development meetings and marketing materials. We've never cared for the phrase because it makes the design process sound as though magical intuition elves are involved, and, like art, you either know how to do it or you don't (that's not true of art, either, by the way). Instead, we like to say *good design*.

Foundational concept: visual space on a page

In Chapter 11 we briefly discussed some design concepts. In this chapter, we explore why those tips work for constructing perceptual explanations in slides, documents, and user interfaces.

For the purposes of design, a page includes three types of elements: black space, gray space, and white space. These element types help you control the perceptual explanations your readers or users create, even if the primary *visual* is text (as in most written documents). Don't forget, though, that pages are found in places beyond written documents.

Black space

Black space is an attention-grabber and doesn't have anything to do with the actual color in that space. Headings, for example, are black space—they draw the eye immediately. Buttons that are a contrasting color (like green) from other buttons can also serve as black space. Pictures or diagrams are black space. Black space *decreases* cognitive load.

Gray space

Gray space conveys the bulk of the information. Text bullets on a slide are gray space. Paragraphs are gray space. Data fields in an interface are gray space. Most of the buttons on a control panel are gray space. Again, this has nothing to do with the actual color of the text or data fields.

Where possible, avoid large blocks of gray space because humans tune out gray space unless they need the information—and most people have a short attention span for consuming gray space, even if they need the information. Gray space *increases* cognitive load because your audience has to perceive *and* process the information.

White space

So, white space is just what's left of the page space after you've designed your black space and gray space, right? Not so fast.

White space serves several important purposes on a page. For one thing, it establishes boundaries around the other elements and provides context (more on that in the section titled "Principle 1: perception is active, fast, and largely preattentive"). For another, it provides *brain rest* and limits cognitive overload. Thus, white space *decreases* cognitive load. Again, white space has nothing to do with the actual color of the space.

Impact on your designs

The good news is you can control how you use space to drive perceptual explanations. For example:

- In **written documents,** the section titled "Building sentences and paragraphs" in Chapter 2 can help you corral visual space into more consumable chunks that are less likely to be tuned out due to sensory adaptation:
  - ▸ The heading writing guidelines discussed in the section titled "Headings" in Chapter 2 help you define black space in written documents.
  - ▸ The sentence and paragraph guidelines discussed in the section titled "Building sentences and paragraphs" in Chapter 2 help you control gray space in written documents.
  - ▸ The space between your short paragraphs and the space around your (ideally short) headings is white space. The top/bottom, left/right margins are also white space.
- In **slides,** Chapter 11 provides guidelines for the amount and type of content that make each slide in your deck understandable and easy to consume.
- In **interfaces,** you can combine the slide design principles noted above with the visual gestalt principles discussed below to organize each UI page and thus guide the entire user experience through your product.

Balance black space (to grab their attention), gray space (to convey information), and white space (to separate elements and make them easier to see). By doing so, you can create at least the basics of strong information design and improve your perceptual explanations.

The next level: visual gestalt

Good design is a learnable skill, and many of the things we cover in this text serve as the foundation for good, intuitive design for pages of all kinds. One of those things is a long-standing set of design principles called *visual gestalt*.[7] Experts disagree about how many principles visual gestalt contains. In this text, we cover what we think are the three most important visual gestalt principles for beginning designers.

---

[7] "Design Principles" (Bradley 2014)

Principle 1: perception is active, fast, and largely preattentive

People actively and unconsciously (preattentively, even) organize the things they see. They make use of everything around them to help them construct meaning. This is true of pages as well.

For example, if you're just sitting in a classroom, it may seem as though you are not very active, right? But your brain is busy constructing meaning for your surroundings whether you're consciously trying to or not. It just happens. From the walls, you perceive that you're inside. From the seats, you perceive that you can and probably should sit. From the presence of a flat surface, you perceive a place to set things down (perhaps the notebook in which you will take class notes, hint, hint). None of this is happening consciously or attentively; it is preattentive.

Further, emotions precede interpretation. Emotions are controlled by much older, evolutionarily speaking, portions of human brains. They help define the fight-or-flight reflex, which is part of how humans are still around today. If people couldn't rapidly perceive the difference between a stalking leopard and waving grass, our ancestors would have been eaten a long time ago.

Humans form impressions based on emotion first (in about 50ms) and rationalize them later.[8] That means that initial impressions are resilient, sometimes even in face of later (better) evidence. So it behooves page designers to practice good information design. If pages are too complex, unattractive, or poorly organized, those making use of them see the information, product, website, and so on as "bad" or "hard"—and that impression lasts a long time even if the item in question is *not* difficult to use.

Using Principle 1 when designing perceptual explanations

What this means for a communicator is that you must control the signal-to-noise ratio in your pages. Effective communication has a high signal-to-noise ratio; that is, you want to reduce extraneous materials or details (noise) and increase pertinent materials (signal). Conversely, poor communication has a high noise-to-signal ratio. Extraneous design elements, poor word choice, distracting colors, and unnecessary detail create a lot of noise.

The fun part is that what is signal and what is noise changes depending on your audience. If you're communicating to a business person, deep technical details are noise, while business-focused material (costs, savings, revenue, and so on) are signal. If you're communicating to other engineers, the business aspects may be noise while the deep technical details are signal.

---

[8] "Study reveals just how quickly we form a first impression" (Dolan 2017)

This is where headings from the writing guidelines help you. These black-space items are always signal. They allow you to identify topics, and they help your readers or users locate the gray space containing the information or active interface segments they need at that moment.

Remember the discussions of color in Chapter 11 and Chapter 13? Colors can influence emotion. If you do not consciously use color to influence your audience, you are inadvertently causing uninfluenced emotions. Uninfluenced emotions are usually noise, whereas the emotion you influence with conscious control of color usually contributes greatly to signal. *Whether emotion is influenced or uninfluenced is entirely up to the decisions you make.*

Principle 2: figure and ground

One of the ways people organize what they see is into figure and ground. *Figure* means the item (paragraph, image, button, and so on) that attracts attention. *Ground* means the items surrounding the figure that provide perspective and context.

Further, the perception of figure and ground is a moveable feast. In a web browser, for example, the address bar is figure when you're trying to find a page, but after you've found that page (taken an action), the address bar recedes into ground, and the page itself becomes figure. And within that page, certain elements can be figure while others are ground until needed.

Using Principle 2 when designing perceptual explanations

As a designer, you can control the perception of figure and ground. To control the perception of figure and ground, you can use black space to draw attention to something you want readers or users to see as figure. Your reader or user will most likely look at the black space first to locate themselves in the document or interface, then move on to looking at the grey space to gain the rest of the information they need.

For example, perhaps the action button in a form in your interface is visually different until all the required information is present. Perhaps you make the starting point or most common initial action more visually prominent. How you do this depends on a conscious use of core page-space principles discussed at the beginning of this section.

For example, if you read comic books, graphic novels, or manga (all examples of perceptually explained stories), you've probably noticed that some areas of the image draw your eye immediately. Maybe the image is darker in an area, or there's a color that contrasts sharply with the rest of the panel. Maybe part of the image is bigger—more prominent due to a controlled use of perspective. That's the artist *explaining* how to read the image using figure and ground.

You can use these same techniques to construct perceptual explanations via your pages. Think about a TV remote: perhaps you make the **ON** button green and larger to use black space to draw the user's attention because that's the first thing they need. Perhaps you set it off by itself so that the white space you designed makes it impossible to miss the black space of the larger or differently colored button. You're explaining how to *see* that button—making it figure.

Think about a software interface or digital touch-control panel: maybe an action button is visually different when inactive (gray space) than when active (black space). If you change a button's appearance after the user enters all the needed information in a form to signal that the button is ready to be clicked or tapped, you're explaining how to use your interface page.

After they're through with the action, your users automatically move that figure to ground, and ignore it until they need it again.

Principle 3: grouping

People group the things into *similar* and *not similar*. In evolutionary terms, for example, humans learned to group "round red things" into the food group (figure) and "flat green things" into the not-food group (ground). But this is a moveable feast (excuse the pun): if what you want is a salad, "flat green things" moves into the food group and is figure, while "round red things" moves into the not-food group (or, at least, "not food I want right now" group) and is ground.

As a designer of pages, you have a fair amount of control over this. If you give unclear clues (round red things are food, except *this* one, which is not, for some unknown and unknowable reason, food), you've provided a bad perceptual explanation and those clues become noise instead of signal. Users then create bad groups because they're *going* to create meaning from your page. If they create bad groups, they do not acquire the correct explanation and struggle to use your product, understand your idea, and so on. And that's on you.

Good grouping aids several cognitive actions that make a page easier to use:

- Scanning to locate the exact information or action needed
- Remembering the location of pertinent elements
- Avoiding errors or mistakes

Using Principle 3 when designing perceptual explanations

Grouping is where black space and white space combine forces to dramatically increase the signal of your perceptual explanation. Black space draws attention to the pertinent actions or information. "Hey, lookit me, right here!" it says. White space around the black space then further frames the black space. *Then* white space around the gray space of the group (related buttons, fields, or paragraphs) says "Don't look past this area, what you want is all here."

Thoughtful, conscious, and consistent grouping increases the signal-to-noise ratio and reduces cognitive load. If you're not consuming all their cognitive load with ineffective perceptual explanations, more of their brain is available for more complex, cognitively constructed explanations.

Document to the question: cognitively constructed explanations

Most software is run by confused users acting on incorrect and incomplete information, doing things the designer never expected.
—*Paul Heckel, The Elements of Friendly Software Design (Heckel 1991)*

If you watch people design or use software (or anything, but our primary experience is with software), often the first thing you notice is that everyone has a different approach to designing, learning, and using each program feature. Given this individuality, how can you ever hope to create interfaces, specifications, and user manuals that more than one person can use?

The best way we've found is to observe people using the product or performing the tasks the software is intended to replace or facilitate. The more you watch them, the more you see patterns emerge that enable you to group information usefully. You may also notice the only constant we've been able to discover: people turn to the online help or paper manual only when they have a specific question, usually one that they cannot answer by experimentation.

By contrast, developers turn to the functional and technical specifications more regularly. However, this does not mean they don't have questions. They do, and your specifications should acknowledge and address this fact.

An excellent paper by Brenda Laurel[9] addresses this issue in terms of online help. And in their book, *Building User-centered On-line Help* (Sellen 1990), Abigail Sellen and Anne Nicol discuss some reasons why users don't like to use online help. They discovered two key points:

- The user's idea of the problem is often very different from the help or program designer's.
- The online help topics were often designed to match the designer's conception, not the user's.

They use these points to group user questions into five categories and suggest that different help content be designed for each of the categories.

After reading Laurel's article, we decided to broaden the ideas from Sellen and Nicol's book to include paper and online documentation as well as content in a product user interface. You can also apply this concept to create usable specification documents and user interfaces.

One of the best ways to corral this idiosyncratic approach and design products with consistent interfaces and behaviors is to create specifications and standards that answer the questions readers have. Although readers may have innumerable individual questions, we were able to boil those questions from Sellen & Nicol's five categories to four basic user questions.

The four user questions

The answers to these four questions include most of the information designers and users seek.

(i) You can't pick and choose which questions to answer in a document, you need to answer them all. And the four questions don't vary based on what type of document you are creating. Readers construct cognitive explanations using answers to *all* of these questions. A best practice is to include at least one sentence in each section of your document addressing each of these questions.

Why should I care?

In the business world, software is frequently imposed from above, to solve problems that management perceives. The people using the software directly may not be consulted—they may just be handed the program and informed that they will now use it.

In this situation, they ask "Why should I care?" or "What's in it for me?" Even if the user purchased or requested the software directly, he or she may only have a marketing-brochure idea of why

---

[9] *The Art of Human-computer Interface Design* (Laurel 1990)

the software is useful. Specific reasons for using specific features to accomplish specific tasks or goals require further explanation. And remember from Chapter 2, this needs to be stated from the point of view of the user, not the product.

In specifications (and we discuss this in much more detail in Chapter 16, *Writing Functional Specifications*), designers need to understand why the feature or function in question is useful—why does the user care about having it in the product?

User guides, overviews, summaries, and theoretical background documents also answer this kind of question. For example, say you're writing a document describing database queries. Information answering "Why do I care" questions includes:

- a brief overview of the process of creating a query and the kinds of results the user can expect
- background on the way tables behave in a relational database structure
- theoretical information on the types of table joins
- a business case in which data gleaned from a well-constructed query saved time and money

The idea is that users want to know why they need to use a software feature. Occasionally, they'll also want to know why they need to use that feature in the specific way recommended by the manual or dictated by the software design.

In specification documents, you can address this question by including a section in the feature or function that explains why the user needs the feature or function.

What is it?
Answering this question just requires a simple description. No matter how intuitive the designer, or even most users, think a product is, there are always users who require more information on its components.

Information answering "What is it?" includes:

- descriptions of the information or choices required by elements of a dialog box or menu in the manual or help file
- glossary definitions of terms used in other descriptions
- bubble help, tooltips, or graphic call-outs describing icons or toolbar buttons

Many users won't push a button, select a menu option, or change program preference defaults until they know what the item is or what it controls.

In a specification, answer this question with a short, specific description of what the feature or function is. Not just what it *does*, what it *is*. Is it a dialog box? A button on a control panel? A connectivity setting?

How do I do it?

This question calls for a procedure. No matter how carefully the workflow is designed, there will be some people to whom it is not obvious. If your product is a general-use computer program (like a word processor, spreadsheet, or photo editor), this likelihood increases. (No two writers or coders approach creating a document or developing code in exactly the same way.)

In a user guide, information answering "How do I do it?" includes:

- step-by-step instructions, using the methods and standards from Chapter 3
- notes, warnings, and cautionary statements

Until users are comfortable with a product's procedures, they won't like using the product. Well-written procedures, following the guidelines described in Chapter 3, make it easier for users to get comfortable using your product. After users are comfortable, routine tasks become preattentive, and they see your product as easy to use because they don't have to read the procedures anymore.

However, after your users are comfortable, you need to be careful and communicate clearly when you change an existing procedure. If you change a feature or function without announcing the change, even experienced users will be unhappy because those comfortable, preattentive tasks no longer work as they expected. You've made their schema wrong: a preattentive activity has returned to being attentive. Your users need to invest time and effort to become comfortable enough with the new procedure so that the activity is preattentive once again.

In addition, you need to be clear when you change an existing procedure. If you change a feature or function without announcing the change, even experienced users will be unhappy because their schema is now wrong. A preattentive activity has become attentive again and will take time to once again become preattentive.

In a specification document, you can answer this question in two ways:

- a short overview of how you, as designers, are going to make the feature or function (emphasis on *short*; this is not the place for deep detail or pseudocode)
- an *action/result* table detailing the user experience (for functional specs) or internal processes (for technical specs) for the feature or function

Why did it do that?

Any object, whether it's software or a telephone or a microwave oven, occasionally displays unexpected behavior. The unexpected is disturbing to users, no matter what level of sophistication they possess. And what you think is fine may be unexpected and upsetting to your users.

If a behavior is weird, but normal, they need to be reassured. If a behavior is the result of an error (either their own or a bug—um, we mean, "undocumented feature"), they need to be reassured and told how to fix it, if possible.

In product development and in user guides, information answering "Why did it do that?" includes:

- helpful error message text, possibly written in passive voice so as to not blame the user
- notes in procedures that anticipate and explain what users may find unexpected
- instructions for correcting the problem, if it is a problem

As with describing procedures (How do I do it?), user comfort level is a serious issue. Many naive users are convinced they have broken something when they receive an error message or a process takes longer than they expected. They need to hear that everything's all right or be given instructions on how to make it so.

In a specifications document, answer this question by anticipating how a feature or function could break, be misunderstood, or malfunction. Then explain how the team is designing the feature or function to prevent, mitigate, or gracefully fail in this situation.

User questions in action

It may seem like you'd be adding a ton of extra information by answering these questions, but in fact, answering these questions is at the core of writing good procedures. Let's look at some annotated examples.

User assistance examples

Let's look at the example procedures from Chapter 3 in the context of the four questions:

> **Adding Contacts**
>
> [*What is it?*] Adding contacts allows you to send emails and other communications to your contacts. [*Why do I care?*] Adding contacts is how people who work in your company are listed in your company directory for everyone to send email.
>
> **Note**: [*What can go wrong*]Before you add contacts, make sure you have added child companies first. When you add a contact, you must assign them to a child company already listed in the system.
>
> 1. [*How do I do it?*]On your dashboard, open the **Contacts** view.
> 2. Click the **Add Contacts** button. The **Add Contacts** screen appears.
> 3. *And so on...*

This example, very simply and easily, answers all four user questions. It doesn't take a lot of reading or complex vocabulary to explain the following to your users:

- what adding contacts is,
- why they want to add contacts,
- what might go wrong while they are adding contacts, and
- how to add contacts.

Updating the BlobBlob

[What is it? and a little Why do I care?] The BlobBlob machine needs to be updated to ensure it has the latest released software. *[Why do I care?]* After you update the BlobBlob machine, you can decode your newest cereal prize. *[What is it?]* When you start an update, the BlobBlob machine runs a system check and updates any out-of-date software in the system. *[What can go wrong?]* Before you start, make sure the BlobBlob is connected to a high-speed internet connection. We recommend a high-speed, low-traffic wireless connection or a T1 line.

> *[What can go wrong? and Why do I care?]* While the update is running, do not disconnect power to the BlobBlob. If the BlobBlob loses power during the update, the update may not complete and data loss can occur.

1. *[How do I do it?]* From the **Home** screen on the BlobBlob machine, press the **Right arrow** 3 times. Press **Enter**. The **Update** screen appears.
2. Press the **Down arrow** 2 times and select the **Update** option. The **Update** screen appears.
3. *And so on…*

This example is a little more complex. Updating the BlobBlob is a more complicated function, so we need to spend a few extra words explaining it. Pay attention, though, to the following:

- Nothing in this procedure violates the guidelines in Chapter 2. Even with the added information, our introduction paragraph is five sentences. All the sentences are well under 25 words.
- This procedure follows the general structure recommended in Chapter 3.
- You don't have to answer the questions in the order we discuss them in this chapter.
- You can combine the answers to questions in one sentence if you are careful and thoughtful—and if doing so doesn't make the sentence too long.

Functional spec example

Let's look at a functional spec example. We'll discuss the structure of this in more detail in Chapter 16, *Writing Functional Specifications*. Here we want you to pay attention to how the explanation is constructed slightly differently for a different audience.

1.1 Mouth Motor

The mouth motor is part of the singing assembly.

1.1.1 Description and Priority

Priority: 1

What is it? The mouth motor controls opening and closing the bottom jaw.

Why Care? Tiffani wants the goat to look like it's singing. She doesn't have any other goat statues or toys that sing, and because she likes show tunes, a singing goat is appealing. She also wants to teach her grandchildren about the musical show tunes she loves.

How Do We Design It? The bottom jaw is hinged, while the top jaw remains stationary. The motor must move the bottom half of the jaw and close it without endangering Tiffani or her grandchildren.

What can go wrong? The closing action on the jaw hinge might snap shut too hard on Tiffani's finger. To prevent this, we configure the hinge to only close 95% with no more than 8 psi.

1.1.2 Stimulus/Response Sequences

Stimulus	Responses
1. Tiffani presses **Start**.	1. The **Start** toggle button sends a signal to the jaw motor at the appropriate part of song playback (see section 4.5) 2. Jaw opens and closes according to signals from playback module. 3. Jaw closes and does not open again at end of song.
4. Tiffani presses Start again or moves power switch to **OFF**.	4. The Start toggle button sends a signal to close jaw and not reopen.

Note how the user questions are still there, but the answers are now adjusted so that they matter to the audience for a functional spec: your co-workers. We still use clear writing, we still follow the clear writing guidelines. We've just changed the focus to connect better to what these readers need and want.

Test case example

Let's look at a test case example. We'll discuss the structure of this in more detail in Chapter 17, *Testing Your Products*. Here we want you to pay attention again to how the explanation is constructed slightly differently for a different audience.

Test Suite: 100 Mouth Motor

Test Case # name: 100.10019 Jaw Spring Tension

Description:

What being tested: This test ensures that the bottom jaw does not snap shut hard enough to injure Tiffani.

Why being tested: Tiffani does not want her toy goat to hurt her or her grandchildren.

How being tested: The test measures the spring tension controlling the jaw hinge closure.

Test Success Criteria: The jaw hinge exerts no more than 8 psi.

Action	Result
1. Connect assembled jaw unit with jaw closed to psi meter.	1. The psi connection LED turns green and the psi meter reads 0.
2. Supply power to motor, press open button.	2. Jaw hinge opens. Psi meter reads no more than 8psi.
3. Turn off power.	3. Jaw hinge snaps closed. Psi meter never reads more than 8psi, when jaw is closed Psi meter reads no more than 2 psi.

Note how the user questions are still there, but the answers are now adjusted to matter to the audience for a test case: your testers. They need steps written like user assistance, but detailed results like developers. We still use plain language, we still follow the clear writing guidelines. We've just changed the focus to connect better to what our readers need and want.

CHAPTER 15

Personas and Scenarios

This chapter is not meant to be an in-depth discussion of personas and scenarios. The full discussion would make this chapter 100+ pages long—there are entire books on the subject. The point is to provide you enough information to understand, create, and use personas and scenarios. To do that, you don't need every painful detail. If you want an in-depth discussion with all the details you might want, see the reference section for this chapter (page 254).

Why do you care?

User-centered product development means understanding who you're developing your product for. This allows you to develop the product your users want, need, and most important, will spend money on. Products that elegantly solve real problems sell better than other products. Elegantly designed products that solve a problem don't happen accidentally.

Personas and scenarios tell you who has the problem and what their life is like. Knowing who has the problem helps you understand how you need to approach solving the problem. Knowing what their life is like tells you a lot about how you need to solve the problem.

You can't develop for yourself or for "everyone"

A common response to "Who are we creating this product for?" is "People like us." This is potentially worse than "We have no idea." It cannot be possible that you're creating a product for people who create this product. Even if you're developing a product for other engineers, those engineers are not like your team of engineers. Your team of engineers is developing this product, which is not what the engineers you're making it for are doing.

Further, as engineers of any type, you are more comfortable using computers than the average person. As Jakob Nielsen, a leading expert on usability engineering (now retired, but his knowledge didn't go away), says: "One of usability's most hard-earned lessons is that you are not the user. This is why it's a disaster to guess at the users' needs. Because designers are so different from the majority of the target audience, it's not just irrelevant what you like or what you think is easy to use—it's often misleading to rely on such personal preferences."[1]

In other words, the user's idea of the problem your product is trying to solve is often very different from your idea of the problem. The technical documents and product designs you create typically reflect your viewpoint, not the reader's.

If you don't know who has the problem you're solving, how can you possibly know what the value of *solved* is for them? Without going through the process in this chapter, you run the risk of developing a product that makes you happy but no one else. That product won't sell very well.

Not everyone wants *your* solution to their problem. In class, Sharon often uses an example of designing a drink that all 100 to 150 people in the class want. How hard can this be, she says? All mammals need to consume liquids or we die, so we know all humans have the problem of thirst. So let's find a liquid everyone in the room likes.

She starts with milk, which she doesn't like, and asks who in the room doesn't like that. She walks through every sort of liquid she can think of—hot tea, cold tea, juices, sodas, wine, beer—and ends with water (which Bonni does not care for). She can't get agreement on any liquid. She can't find any *one* liquid to solve the human problem of thirst that 150 or so, frankly pretty similar, people can all agree they *like*.

If 150 similar people can't come up with any one liquid they all like for an activity they all *must* perform or they die, you're not going to solve a more complex and voluntary problem in a way that makes sense for all humans. "Everyone" is not your market.

What are preferred input modes?

Preferred input modes are, quite simply, the different ways weird and wonderful human brains like to take in information. While everyone is unique, and none of this is locked in concrete, people prefer to take in information in the most comfortable way for them. And because humans are all built from the same building blocks, those ways tend to cluster around human senses.

---

[1] "The Distribution of Users' Computer Skills" (Nielsen 2016). Although this article is from 2016, the statistics he cites haven't changed all that much, and there are newer articles that suggest literacy is dropping in the generations that grew up with smartphones, including: "U.S. Students' Computer Literacy Performance Drops" (Langreo 2024) and "Turn it off and on again" (Zerrenner 2024).

There are many theories about what those groupings are, but the one we think resonates most in the technical communication world is VARK,[2] which identifies four modes by which humans take in information. Those modes are:

- Visual (sight)
- Aural (sound)
- Reading (text)
- Kinesthetic (motion)

How do I improve my communication using VARK?

Many, if not most, people are *multi-modal*. That is, they are equally comfortable taking in information in more than one mode. Strong communication includes elements of all four modes.

Let's explore the modes and methods used to communicate in more detail.

V = Visual

People who favor this input mode prefer a more graphical approach to taking in and delivering information. They tend to sketch out their ideas. If you ask them for directions, they're likely to draw you a map. If they need help with a product, they tend to want videos or plenty of screenshots.

You can most effectively communicate with visual types by including:

- Pictures that *communicate the information*. Just any old picture is not enough.
- Diagrams/charts/graphs (e.g., flowcharts, line graphs, and so on; see the section titled "Graphs and charts" in Chapter 11 for more).
- Color (when used correctly; see the section titled "Colors" in Chapter 11 for more).
- Blinking or color-coded status lights.

A = Aural/Auditory

People who favor this input mode prefer to hear and say information. They tend to talk through ideas in brainstorming sessions. If you ask them for directions, they're likely to tell you. Be prepared to take notes because they won't write them down. If they're seeking help with a product, they ask a co-worker or call the help desk/support line.

---

[2] "VARK" (VARK Learn Limited 2021). Ironically, "digital natives" seem to have lower general computer literacy than "digital immigrants"—those of us who had to acquire our knowledge of what was then "new technology." See Chapter 10, *Pitching Ideas*, for a more detailed discussion of "new technology."

You can most effectively communicate with auditory types by doing the following:

- Writing in a conversational tone so they can *hear* your writing as they're reading[3]
- Including audio prompts, such as a beep, in your product
- Consider a podcast or *talking-head* video

R = Reading

People who favor this input mode prefer to read text on a page or screen. They tend to note down their ideas using words, rather than pictures. If you ask them for directions, they typically carefully write them out and hand you the piece of paper. If they're seeking help with a product, they read the manual or help file or search the internet for articles and knowledge bases.

You can most effectively communicate with reading types by following the writing guidelines discussed in Chapter 2. Just because they like to take in information using words doesn't mean they want to read an unnecessarily large number of confusingly constructed words.

If you provide a video for your audience, using captions on your video helps reading input mode people. They'll typically turn the sound off and read the captions, often at 1.5 speed.

K = Kinesthetic

People who favor this input mode like to take in and deliver information in a way involving motion or activity. They may sketch out their ideas (drawing is motion), write down their ideas (writing is motion), or talk through them while waving their arms (talking is also motion). If you provide a model for these people, they pick it up and work with it in their hands. If they need to sit in a meeting, they may have fidget toys to handle while they listen.

If you ask for directions, they act out the directions, for example moving their hand to the right to indicate a right turn. If they're seeking help with a product…well, they won't. They usually just try stuff until something works or goes horribly wrong, at which point they fall back to their secondary input mode. As does everyone.

How can you communicate effectively with kinesthetic types? Cut to the chase. Get them into the action as quickly as possible and give them a reason for what they're doing. Write instructions that have their hands on the product as quickly as possible.

---

[3] The writing guidelines from Chapter 2 can help you write in a conversational tone.

Input modes versus learning theory

The input mode theory we discuss in this chapter is often referred to as a learning theory. However, it has been widely discredited as such.[4] Many of the studies that discredit the modes as "learning theory" examined these modes as applied to children in school. In fact, we agree: this theory is not effective when applied to pure learning situations, such as a K-12 classroom.

VARK is slightly more effective when applied to a college classroom or lab. A study in the Iranian Red Crescent Medical Journal[5] talks about learning modes for medical students. In these situations, participants are primed to learn and are more plastic in the modes of communication they accept and adopt—although adult learners still have clearly defined preferences.

However, when applied to adult learners or workplace communication, our experience indicates that the modes are significantly more effective and important. While people can learn and take in information using all modes, as adults they are less plastic and have preferences, and those preferences can be strong. Although these preferences evolve due to experience, people form a schema (in this case about themselves, sometimes unconsciously) and habits around that schema.

Trying to take in information in ways that do not conform to people's preferences can create fierce interference. You add cognitive load to the information by presenting it in a non-preferred input mode, which interferes with understanding. What that means is people just report that something is *too hard*. It's too hard to use your interface. It's too hard to watch the video. It's too hard to read the information. Products that are too hard to use don't sell well.

A wise engineer takes input modes into account when developing personas—and all the design that happens based on those personas.

Why should I care about input modes?

You've watched people to whom you've given careful directions get lost anyway. You've seen texts or Slack messages go awry or get ignored. You've felt the frustration of trying to explain something you feel you're expressing perfectly clearly to someone who seems to be *completely missing the point*. You've struggled to understand something and blamed yourself for just not getting it. You've worked with software interfaces and control panels that seem to make no sense.

---

[4] "The Stubborn Myth of 'Learning Styles'" (Furey 2020)

[5] "The Relationship Between Learning Style Preferences and Gender, Educational Major and Status in First Year Medical Students" (Sarabi-Asiabar 2014)

Why does this happen? Part of the problem is that people tend to discount, ignore, or poorly absorb information that does not arrive via one of their preferred modes.

Humans tend to communicate most naturally in the modes they prefer, which is fine. Except that communication is (or should be) more about the receiver than the sender. Knowing your own preferred input modes helps you study better, communicate better, and design better products.

All parts of a product communicate: the user interface, the control panel, the web-page design, button labels, and so on. The more you understand input modes, the stronger your human-device communication can be, and the better your product can be because it makes sense to your users.

Knowing the input modes of your co-workers and making a strong guess at the preferred modes of your audience builds better communication, which can lead to a better performing team. This helps ensure you're using the best techniques to communicate all those wonderful ideas in your head to create products that truly solve a problem.

Personas

Personas are the mechanism practitioners in engineering and related fields use to humanize their audience. As discussed above, products solve problems. *People* have problems. Therefore, taking the time to understand the people whose problems you're solving results in products that solve those problems more effectively. And, as discussed in Chapter 5, solving problems effectively results in profitable and, more importantly, long-lived companies.

A persona is not a real person. It is a carefully defined *fictional* person who represents a *hypothetical archetype*[6] that:

- includes specific details culled from marketing demographics and other audience research (you'd be surprised how much is freely available on the Internet)
- has specified input modes (when in doubt, employ the tips for all four input modes)
- is given a name and a picture
- is precise, rather than accurate

---

[6] *The Inmates are Running the Asylum* (Cooper 2004)

We cannot stress strongly enough that personas should be fictional. Do not base personas on any given real person. Real people have idiosyncrasies that can badly skew how you approach communication. This is what we mean by "is precise, rather than accurate"—your persona should include details to make it *seem* like a real person without actually *being* a real person.

How do I create a persona?

Start with your marketing research. What are the basic demographics of your primary audience? Think:

- **Gender:** although gender rarely has an effect on most products, there are some key exceptions, so be alert for those
- **Ethnicity:** that is, include some diversity; your audience is highly unlikely to be a monolithic block of a single ethnicity
- **Age:** because, for example, 6-point font without contrast is bad if your audience is people over 40 years of age
- **Education:** although this may not be an issue, if your product is for CPAs, for example, they are experts in the accounting domain of finances
- **Income:** because at the very least, whether they can afford the product is a consideration,
- and so on....

Choose demographics from the mid-point of the bell curve—you're trying to describe the average user, so go for the average data. Table 15.1 is an example of demographics for a persona.

Table 15.1 – Sample demographics for a persona

If your audience data is	Your persona becomes
Age 25–45	32-years old
College graduates	BA in Computer Engineering from the University of CA, Riverside
Upper-middle income	$150,000/year

However, don't rely on simple demographics: also include details relevant to your product. Demographics alone do not tell the whole story, as you can see in Figure 15.1.[7]

Edward Bell
Head of Digital | Marketing Leader | Digital & Data Expert |
Father of three

If you were segmenting based on demographics alone, these two would be in the same target audience. Just something to think about…

Prince Charles

Male
Born in 1948
Raised in the UK
Married Twice
Lives in a castle
Wealthy and Famous

Ozzy Osbourne

Male
Born in 1948
Raised in the UK
Married Twice
Lives in a castle
Wealthy and Famous

Figure 15.1 – Demographics can mislead

If you're developing, say, a game or device to help someone improve their golf game, "likes golf, but is a terrible player" or something similar should appear in the persona. Include as many of these details as you can. The better you understand the experience, habits, and schema your persona embodies, the better your product helps them solve their problem.

How do I use a persona?

The persona guides product development. Plan to refer to the persona constantly when developing and defining the features and functions of your product. Persona details drive:

- which features and functions you include to have a product that solves their problem
- what those features and functions are and how they work
- what constraints the persona's characteristics introduce

---

[7] Recreated from Edward Bell's LinkedIn page (Bell 2021) using available images.

Scenarios

After you have at least one rich and detailed persona, you know who you're making the product for. The next step is to understand and define what their life is like. This is called a *scenario*. Every persona should have at least one rich and detailed scenario.

A scenario is a detailed story about your persona's life, such as a day at work, or other relevant part of their life. Scenarios do not have the persona interacting with your product, nor does the product swoop in at the end and save the day for your persona. Scenarios are a detailed description of the motivations and concerns the persona faces on the job or in life. They provide the *context of use* for your product.

In an agile environment, these can be thought of, and are often used as, *user stories*. They detail how the persona's life works and can help you understand how the product can fit into that life. Regardless of your development environment, scenarios are not like an advertisement where the persona has the product and now they have a perfect life.

Scenario example

Mary has a busy life, chasing after three kids under eight. Two of them are in school and after the chaos of getting them out the door and at school, her day calms down a bit. Now it's just household tasks, like buying groceries, laundry, and picking up the always messy house.

However, the 3-year-old is starting to not want naps in the afternoon anymore. When it's time for the older kids to get out of school, they go to sports and dance class. Then it's pick them up, get everyone home, and make dinner while she directs the children on their homework and keeps them from fighting before she can feed them.

Her husband, John, gets home in time for dinner, and he takes over with the kids, bathing and getting each kid to bed. Mary cleans up the kitchen, makes lunch for the kids for tomorrow morning. They both collapse on the sofa for an hour of staring at the TV and catching up on their day before they go to bed.

If they're lucky, the youngest will be up only one time during the night. They haven't been very lucky lately.

This scenario mentions almost no products, certainly nothing by name brand. This scenario works for at least the following products:

- Electronic tablet
- Thermal drink cup
- Television
- Dinnerware
- Car
- Dining table
- Household cleaner
- Frozen meals
- Snack bowls
- And many more…

Notice the scenario doesn't have Mary or her husband wishing for your product to make their life easier. We especially don't have them using your product, and now their life is wonderful. That's not the point to a scenario—scenarios show you what their life is like, not what their life would be like if they were using your product. Scenarios show the *operating environment* your product must function in.

Note, however, that including something in their day that demonstrates that all is not going well in some area (ideally the area that your product can affect), helps provide additional context for your scenario.

Understanding what their life is like lets you know what the operating environment is like. Using the example scenario, you know the operating environment must account for three active small children with involved and tired parents.

These children participate in after-school activities. Travel bags of equipment are stored in the car and removed for the activity, probably by Mary. She probably delivers the bag and the child to the leader of the activity and then gets back in the car to get the next child to their activity. You know the children are also picked up and brought home, where they do their homework and quarrel. You also know the parents are tired and may not attend to every detail in their lives.

Regardless of the product you create, you need to account for these considerations. For example, a fragile dinner set will not be interesting to this family. Any storage product for the car needs to be easy to get equipment in and out of and large enough to store at least two bags of kids' activity equipment. Any transportation technology must account for five people, including three children and their booster seats. And so on.

Requirements are implied

None of this is mentioned in the scenario; it's implied. You might list these and more after the scenario as environmental concerns (or as product-relevant characteristics in the persona), but you infer these from the scenario. Your job as an engineer is to make these inferences and then write them down to share with others on the team. As you build the scenarios, these inferences are made explicit for confirmation with the other people working on the project.

In an agile environment, scenarios are called *user stories*,[8] but regardless of the development environment—and regardless of whether your company calls them user stories, scenarios, or something else—they feed product development.

Developing scenarios helps in product development because it can shut down feature creep. When someone wants to add a new feature, you can ask if Mary and Mary's life need that feature in the product. The answer might be yes, that's a useful feature, but more often, the feature doesn't fit Mary's life and the problem you're trying to solve.

Scenarios do not include the product

Scenarios do not include the product because we don't want to know how Mary's life is now wonderful because she has the product. That provides us no information that's useful in *designing* the product to solve Mary's problem in a way that fits in her life. We need to know the environmental ecosystem of her life to design a product that works in her environment *and* solves a problem she has.

For example, Mary most certainly has the problem of thirst. She's a mammal and mammals have that problem. We could design a lovely hand-blown glass to-go cup that fits nicely in her car cup holders. That solves a problem Mary has—she gets thirsty and is in the car a lot. We could pat ourselves on the back for a job well done.

---

[8] Sometimes scenarios are in the *epic* (a bigger collection of user stories) because they can be big. Smaller stories that are related to the scenarios are the user stories.

But Mary will never buy our product. We've ignored the full scenario—the environmental concerns of her life. Yes, she is in her car a lot. But we ignored the part where she has three active children she transports to and from school and activities.

Children grab parents' drink cups. Children grab things from other children's hands. Children drop things. Mary, getting activity bags in and out of the car with her keys in one hand and the other holding a child's hand, drops things. Our lovely and fragile to-go cup wouldn't last a day in her life. She would never buy this product because it would soon be just pretty pieces of broken glass she has to clean up.

There may be a persona and scenario for someone who wants our lovely hand-blown glass to-go cup. Mary is not that persona, and her life is not that scenario.

Wrong personas and wrong scenarios

It can be tempting to ignore the persona and scenario process because you just want to get started on developing the product. You may think this is all just overhead that's not needed and gets in the way of product development. But you would be wrong.

Defining who has the problem you're solving and what context the product will be used in is, in fact, starting to develop the product. It focuses the team on what you're developing and why you're developing that specific feature set. It exposes requirements you may not otherwise have thought of until far into the product development process, where it costs a great deal of money to change. Or you may never have thought of the requirements at all.

That said, you can't just make up personas and scenarios and think you've got it. Both of us have seen companies invest a lot of time and money making detailed personas and scenarios, and then build and deliver products that were a crashing loss in the marketplace. Often, the company used the wrong personas and scenarios, and the sales reflected that.

Learn from these companies and internally socialize your personas and scenarios when you think you're done. Think about the flow of a project and talk to other teams to see if these make sense for them. Reach out to Sales and Marketing to find out if what you developed makes sense to the people who are marketing and selling the product.

What do you do with personas and scenarios?

Don't spend months creating personas and scenarios and then shove them in a drawer, never to be thought of again. Post them in a common area where team members see them every day.

Refer to them when you're making design decisions. Ask yourself: is this how Mary would expect this to work? Does this fit in Mary's life, or do we need to change it? When people want to add or remove features, refer to the personas and scenarios for guidance. If you add the feature, what needs to happen to fit into Mary's life? What is the minimum viable product that solves Mary's problem and fits into her life? What priority is this feature to Mary?

When the reality check meeting happens, the personas and scenarios should feed into the features that get cut. If you cut the features that matter most to how Mary wants the problem solved, there may be no point to delivering the product at all.

CHAPTER 16

Writing Functional Specifications

At some point, you need to write down what it is you want to develop, regardless of the product development method your company uses. Many companies use a development group that works in a different location than where you are located. This is even more true if your company supports remote work. The details of what you want to build need to be communicated to them.

Even in an agile environment, you must communicate what is being built in each sprint. Often, this is through a functional specification of some sort.[1] A hardware environment typically uses a fully developed functional spec because it's hard to do hardware in sprints. Not impossible, because no one said you must ship the product at the end of each sprint, but it is harder.

In this chapter, we make an artificial distinction between features and functions. You will never see this distinction in the business world. We make this distinction because the way you approach each is slightly different.

- **Function:** a function is a *behavior* of the product, a thing the product *does*. Think verbs, such as run a report, transfer files, increase or decrease volume, and so on. For example, "log in" is a function—it is an action a person or a part of the product must take to achieve product goals. Functions are active.
- **Feature:** a feature is a *quality* of a product, a thing the product *is*. Think adjectives, such as rugged, cute, easy to use, and so on. For example, "simple login" is a feature—it is a *characteristic* of a log-in function. Features are passive.

If products solve problems, then each feature or function either solves part of the problem or is a necessary step in, or attribute of, the solution. Designing and specifying features and functions is a great place to rely on your personas because they have the problems you're trying to solve.

Note that this is a distinction you may only see in this text. Most companies use these terms interchangeably—and, in fact, some methods of creating a spec blend behavior and characteristics in the description. We draw a distinction here because the way to approach features vs. functions in a functional specification is similar in structure but different in implementation.

---

[1] In an agile environment, the functional spec for a sprint is often represented as a burn-down list or chart, often in a product like Jira, which shows the amount of work left to do in the sprint.

The nature of a functional specification

There is no industry standard for what a functional specification looks like or what's included. Your company usually has a structure they use, and they like it. Use that structure. Here we discuss what goes in a functional specification in general, based on the IEEE-suggested template.[2]

Before exploring the functional specification, let's discuss the different types of documents you may create in a product development environment and identify which are truly specifications.

User assistance

User assistance is the information, or content, that instructs and supports a user as they use the product. This content includes some or all of the following: a printed user guide, online help, a knowledge base, or words on the interface. For ease of discussion here, we're going to use the phrase *user assistance* to refer to all this sort of content.

User assistance is not a specification. While it is indeed content that describes specific actions users must take to accomplish a task (see the section titled "Document to the question: cognitively constructed explanations" in Chapter 14, *Constructing Explanations*), this document type is distributed outside your company. External documents do not and should not explore the same level of technical detail as internal documents do.

User assistance explains exactly how your customers can accomplish their goals or solve their problem using your product. User assistance is written for your persona so that they can understand the steps of how to use the product to accomplish the tasks they need to accomplish in your product. Think back to Chapter 3, *Writing Good Procedures*.

> (i) User assistance is not written for everyone. "Everyone" is *never* your product user, and that's even more the case for the user assistance content because some users will never access the user assistance. For example, your users may be nuclear physicists with a sophisticated and nuanced understanding of thermonuclear reactors. You never provide these personas with user assistance "everyone" can use. You provide persona-appropriate information for the complex product you are creating.

---

[2] *IEEE Software Requirements Template* (Wiegers 1999)

Technical specifications

At the opposite end of the spectrum, the technical spec describes functional behavior and detailed characteristics focused on *machine/internal expectations and needs*. It is a very protected, internal document, typically read only by engineering team members.

The technical specification provides deep details about the underlying technology of a product. It often contains company secrets and may include ideas and techniques the company can patent.

This document is written expecting that the audience is expert in making these products because the audience is your engineering staff. Clear writing is important because the information in this document is typically the only information the product development team has to know what they are building. Your engineering team may be 12-hours offset from the development team that is using your technical specification to build the product. It's hard to have casual conversations about the technical specification with that large a time difference.

Functional specifications

Square in the middle between user guides and technical specifications lies the functional specification. Also a protected and confidential internal document, the functional specification describes functional behavior and high-level characteristics of a product in terms of features and functions. Heavily based on your personas, the functional specification focuses on *persona (human) expectations and behavior*.

This document is written so that the people involved in building the product can understand what's needed to get this project completed. Think back to Chapter 9, *Flow of a Project in a Company*, where we discussed the many roles in a company. Any of those people may want or need to read a functional specification.

Deconstructing functions and features

What you think of as a function or feature is probably actually a collection of functions, features, or both—in other words, a *system*. In a specification, you have to unpack, or deconstruct, these systems into smaller parts. Often, this deconstruction in the functional spec generates multiple technical specs.

Let's take a look at the example of secure login. We can deconstruct that into the following elements, each of which must be specified:

- The encryption method (feature)
- The encryption itself (function—machine)
- The user log-in screen (function—user)
- The authentication (function—machine/background)
- Post-authentication (function—machine)
- The ease of use (feature)

Each of these must be documented separately because:

- They generate different questions and represent different characteristics or behavior.
- It's possible that different teams are working on each element, and you don't want to make each team wade through a long document to find the one piece of information it wants.
- They affect different things.

While you can document all of these in a single section (instead of in individual sections for each feature or function listed), it's a bad idea to do so. That section becomes long, complex, and hard to read. In addition, you're muddying the waters by trying to combine a feature (security method) and several functions (the rest of the items) that have different actors and often different systems.

When you muddy the waters in this way, it is easy to forget or omit important details that can cause difficulties or rework later in the development cycle. It's inherently confusing and cognitively expensive for the readers.

Therefore, it makes sense to document the smallest sensible specifiable unit and treat feature or function integration separately. For each of the examples below, note how what looks like a simple device or app actually contains many parts, each of which needs to be described individually, with enough detail to design the part.

Writing functions and features

While different companies use different templates, most include some variation on the elements described below. This section discusses how to document product functions and, where they differ, features.

Priority

It is a sad fact of product development that time is not infinitely flexible. At some point, you have to ship the product, and therefore, you must make hard decisions about what to remove from the development of this version. By setting a priority on each function or feature, from the persona's perspective, these decisions are made more easily in the reality check meeting.

Whatever priority you assign to a feature or function, you must include a legend of the priority scale. Some team members think in terms of lower numbers being higher priority, while others think in the opposite direction. Your entire product team must start with a common definition of priority and document that clearly in the functional specification.

Multiple features or functions can have the same priority from the scale. For example, many features or functions may have a priority of 1, the highest priority. That rating means these features and functions must be in the product for the product to be the Minimum Viable Product (MVP).

Other features and functions may have a priority of 5. Assuming our scale goes from 1 to 5, that means that if we have time, features and functions with a 5 priority go in the product. They are nice-to-have features and functions, but not critical.

Description

This section provides an overview of the function in question. Remember the four questions discussed in the section titled "Document to the question: cognitively constructed explanations," in Chapter 14, *Constructing Explanations*? A good description section answers these four questions, although it approaches them slightly differently from a user guide.

What is it?

To answer this question, be literal. Is the function software? hardware? an interface item? a programming function? Summarize this answer here—you can go into deeper detail in the associated technical specification.

For example, in our secure login user dialog function, we can use the following to answer this question:

- The user dialog is an interface with two fields and one action button:
    - Username field (up to X alphanumeric characters, excluding special characters not typically found on a standard personal computer keyboard)
    - Password (up to Y alphanumeric and special characters, excluding special characters not typically found on a standard personal computer keyboard)
    - **Log In** button (the entire button is clickable only after both text fields have information typed in them)

For the ease-of-use feature, we can answer this question as:

- The login screen contains only the information that Marissa needs to log in.

Why does [*persona*] want this feature?

To answer this question, you must review your personas, identify one (ideally the primary persona), and put yourself in their shoes. In this example, and continuing into the next chapter, we'll call our primary persona Marissa. Why does this person need the function? What part of the overall product problem does it solve? We can use the following to answer this question for both feature and function:

- Marissa needs to securely and easily access her bank account, ensuring that she is the only one who can access her account.

Note this section is not why the persona wants the entire product—rather it's why the persona wants *this* feature or function specifically.

How do we design this?

To answer this question, summarize the general code or design the development team plans to use to solve this part of the overall problem. How are you going to implement this feature or function? What's going on that your persona *doesn't* see? Don't go into deep technical detail here—that's for the associated technical spec. Just provide enough of a summary to let the technical spec engineers get started.

Continuing our function example further, we can use the following to answer this question:

> - We create a dialog box containing three elements as noted above. When Marissa clicks **Log In**, we validate the username and password fields against the field criteria. If the entries meet the validation criteria, we pass the values to the authentication engine (*include a cross reference here to the section about the authentication engine*).

To answer this question for a feature ("login is easy to use"), we can say:

> - When Marissa sees the login screen, it contains only the fields and buttons she needs. The **Log In** button is clearly distinct from, for example, the **Create an account** button. The **Log In** button is not active until she enters her username and password, so she cannot take action until all the login information is present.

What can go wrong?

To answer this question, think about what can go wrong with the feature or function you're engineering. After you identify at least one thing (because there's always at least one thing), identify how you prevent or mitigate that issue through engineering.

We know that products can and do fail. A good and ethical product developer prevents as many failures as reasonably possible. In fact, as an engineer, it is your ethical, moral, and sometimes legal responsibility to anticipate and prevent as many failures as reasonably possible.

However, it is not possible to prevent all failures. For example, if Marissa accidentally types a character in her username that's not allowed in that field—that's not under your control. Where you cannot prevent a problem, your product should mitigate it with what's called *graceful failure*. That means you must ensure that the failure of the function is neither catastrophic nor confusing.

In our function example, we do not want the login to lock up if Marissa accidentally types a disallowed character. That's ethically rude. In addition, we don't want to just re-display the log-in screen and not tell Marissa why she's not logged in—she didn't enter the wrong password on purpose. However, someone *else* may be trying to log into Marissa's account. We may not want to provide enough information to allow a potential hacker script to know what the login issue is. We need to specify—to *engineer*—how we're going to handle it.

Finishing our example function description, we can use the following to answer this question:

- Marissa may mistype her password or username using characters that are not allowed for those fields. To mitigate this, we define the fields using coding to immediately verify her entry and show an error message about invalid log in information when she moves to the next field.
- Marissa may accidentally click the **Log In** button before she types her password. We disable the **Log In** button until text is typed in both fields.

For our feature example, we can say:

- When Marissa sees the screen, she could mistype her username or password. To fix this, we include an error message to explain in clear language that her login failed. We also provide general instructions for fixing the problem (for example, check your CAPS key).

Action/result tables

The action/result table details the exact user experience steps on the action side, while summarizing internal product behavior steps on the result side. This table gives your product team a clear idea of the behavior flow and expectations in a step-by-step way, expanding on the user experience summarized above.

Note that the action/result relationship is not necessarily one-to-one. One user action may drive multiple results from the system.

Recall Chapter 3, *Writing Good Procedures*. The same structure of numbered steps applies here as well, but pulled apart for clarity. In this structure, put the action the persona takes on the left side of the table. Put the result of that action on the right side of the table.

An action/result table for our sample function can look as follows.

Action	Result
1. Marissa types her username. 2. Marissa presses TAB to move to the password field.	The Secure Log-In dialog validates Marissa's username against defined field criteria: • If the entry passes validation, the cursor moves to the password field. • If the entry does not pass validation, the dialog clears her entry, leaves the cursor in the username field, and shows an error message describing the issue and how to fix it.
3. Marissa types her password. 4. Marissa clicks **Log In**.	The Secure Log-In dialog validates Marissa's password against defined field criteria: • If the entry passes validation, the default banking screen appears. • If the entry does not pass validation, the dialog retains her username entry, clears her password entry, leaves the cursor in the password field, and shows an error message describing the issue and how to fix it.

An action/result table for our sample feature can look as follows:

Action	Result
1. Marissa accesses the login screen. 2. Marissa sees only the fields and buttons she needs for login and nothing else.	The following items appear on the screen: • Username field • Password • Inactive **Log In** Button

Business and functional requirements

The business and functional requirements define conditions beyond the function itself that must be met for this function to behave as described. This is where you can connect this function to other functions or features. Think through the user scenario you defined for this persona, as well as any additional technical requirements.

To finalize our function example, we can specify the following:

- Base code must accept AJAX field checking.
- Dialog must pass field entries to the authentication engine.
- Dialog must appear when the application launches.
- Field sizes on the dialog must be long enough to display the entire entry.
- Password entry field must obscure characters on screen after entry.
- Password obscuring must wait 1 second for user verification before obscuring.
- System must limit the number of failed attempts to three.

To finalize our feature example, we can specify the following:

- Dialog box must be obviously accessible from the home screen.
- Login page must have responsive design to automatically adapt to the correct screen size for desktop, laptop, and mobile devices.
- Must have a dedicated 24/7 phone number for customer support displayed on the screen.
- Must be coded to support cost-effective localization (for example, store all on-screen text a linked text file).

CHAPTER 17

Testing Your Products

Testing is an essential part of the product development process, regardless of the product development method your company uses. As a former student said, "I now think of testing as a way to honor the trust that users place in the products we design." In other words, robust and thorough testing demonstrates your engineering ethics.

Testing is not only essential, it's morally, ethically, and almost always legally required. Testing is an important thing to get right, and getting it right depends on good communication.

Organizational styles

If you're in a waterfall development environment (more typical for hardware products), testing is a separate and distinct phase that happens after the product is developed or at least large subsections of the product are ready for testing. If you're in an agile environment, testing happens as you develop the product and stories are moved across the board. Much smaller parts of the product are often tested in an agile environment, simply because these parts are available faster.

But regardless, you must test your product. Typically, that means some sort of testing documentation so that people can rigorously test the product. This chapter is about creating those test documents, not about how to test your product.

If testing is interesting to you, there are large international organizations dedicated to industry standards for testing, research into testing, and education for people in the testing world. They are robust organizations with lots of information about the philosophy of testing. The Association for Software Testing (AST), Software Certifications, Sticky Minds, and International Institute for Software Testing are some organizations worth investigating.

This chapter is about how to create test cases that you can use to test products, regardless of the development methodology you're using.

Why test?

In a startup environment, when it's time to test, it's all hands on deck to poke at the product to see if you can break it. Often, everyone in the company is involved in poking at the product in creative and unstructured ways, writing down what they did that made the product behave badly.

This can be really fun. Both of us have done these crash-and-burn testing phases so the product could ship on time. In a startup environment, shipping the product is the most critical thing you can do because venture capital is often riding on meeting the deadline. Sometimes your company's survival is on the line.

But this is not a sustainable testing methodology. Too many things get missed that must be fixed when the product is in the field. Fixing issues after the product is released is the most expensive time to fix them. You also can take a market hit as your industry realizes you released a buggy product. At some point, your company matures enough to require a more formal testing process.

If you're in a regulated industry, such as medical devices, your testing methodology and results are dictated to you by governmental agencies. Even as a startup, you can't ship buggy medical devices. You must have a rigorous and documented testing process. Frequently, you must show your testing process to the oversight agency, including the testing documentation, to be allowed to sell your product.

Further, people tend to be blind to their own mistakes. This applies to you as a product developer and communicator as well. It's very unlikely, for example, that your code compiles on the first try. Even if it does, compilers can only determine whether your code follows the rules of the language you are coding in—it can't determine whether it will run correctly and be free from defects. A perfect compile does not mean your software is defect free.

From a business perspective, defects can cost a company a lot of money, particularly if they are found late in the process. The later they are found, the more expensive they are to fix. This is true even if you're working in an agile environment and release patches and new features frequently and regularly. Think about the flow of a project and what's involved to correct a core issue found just as you create the final build (see Chapter 9, *Flow of a Project in a Company*, and Figure 17.1).

What is a defect?

The Quality Assurance Institute defines a defect as an "undesirable state." The undesirable state can be as simple as a spelling error on an interface page or control panel or as serious as a flight control system malfunctioning, causing a plane to crash (looking at you, Boeing). Regardless of the severity, testing helps you identify and repair defects.

Defects can often be traced back to requirements and specifications. Reasons for this include:

- a lack of clarity in the functional or technical specification (see Chapter 16, *Writing Functional Specifications*, for tips on avoiding unclear specifications)
- errors in the specifications
- ignoring the persona and scenario while developing the functional specification
- not thinking through all the implications of the product, features, or functions

Quality is everyone's job, and it needs to start early in the flow of a project. Given that defects often trace back to specifications and requirements, you and your testing group can (and should!) start testing and planning for testing as soon as you have a credibly complete draft of the functional spec—words are easy, fast, and cheap to fix.

The earlier you can fix a defect, the cheaper it is over the product lifecycle. Figure 17.1 shows an approximation of the cost to repair a defect, based on when it is found. The later you fix a defect the more expensive it is to fix, and after you get to the production phase, the cost can be an order of magnitude higher.

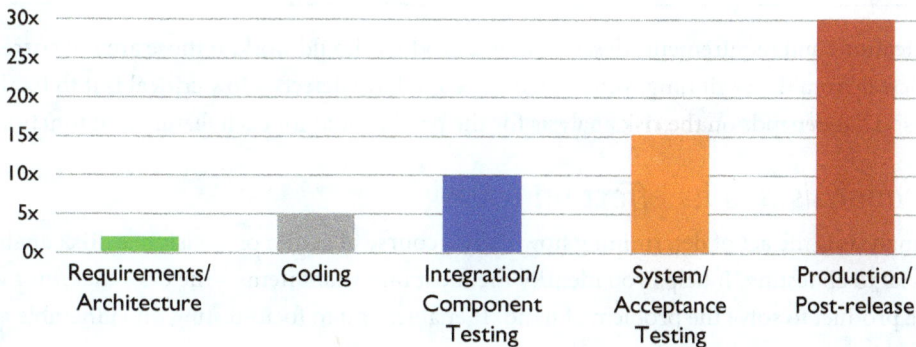

Figure 17.1 – Relative cost to fix defects in each project phase[1]

---

[1] This is an approximation based on several studies cited in "The Economic Impacts of Inadequate Infrastructure for Software Testing" (RTI 2002).

What can be tested?

Effective testing starts at the very beginning of the flow of the project, also known as the product development life cycle. Table 17.1 shows the phases of the lifecycle and what is tested in each phase. This breakdown applies whether you're working in a waterfall or in an agile environment.

Table 17.1 – Testing across the product development lifecycle

Phase	What's Tested
Requirements (functional specifications)	Precision, understandability, completeness of specification.
Design (technical specification)	Precision, understandability, feasibility, completeness of specification.
Development	Core functionality of each unit, some multiple-unit functionality, and usability.
Testing	Expected vs. actual behavior/performance, maximum load, usability, and so on.
Installation	Behavior/performance on user systems.
Maintenance	User-reported problems, integration of new features/functions with existing features/functions, and so on.

Specifications and requirements describe how a product should work. If those are not correct or are unclear from the beginning, you're guaranteed to have defects. How critical is it to fix those defects? That depends on the risk analysis for the product and for each feature and function.

Risk analysis and its effect on testing

Risk analysis is the act of determining how risky a course of action or decision is. Risk analysis is crucial to all testing. It helps you identify the most important items—the ones that *must* work for the product to solve the problem at hand. It enables you to focus testing on vulnerable areas and avoid over- or under-testing.

Calculating risk

There are many methods for calculating risk, some more complex than others. We recommend the following simple version during functional specification design and initial test planning:

1. Assign a number from 1–9 for the *likelihood* of failure.

   Factors affecting likelihood include:

   - **Complexity:** the greater the complexity, the higher the likelihood of failure
   - **Scale:** the larger the scale, whether of the feature/function itself or the amount of usage, the higher the likelihood of failure
   - **Age/stability of technology involved:** the newer the technology, the more likely it is to be unstable, at least at first, and the higher the likelihood of failure
   - **Experience with colleagues:** the less experience you have working with the team, the more likely you must work out communication problems and/or trust issues, and the greater the likelihood of failure

   Where you physically work is typically not a risk factor (assuming you have reasonably stable power and internet). If you or anyone on your team works from home, that does not increase risk. Working with an overseas team *may* increase risk, but that risk can usually be controlled with the type of effective and complete communication we teach you in this book.

2. Assign a number from 1–9 for the *severity of consequences* of failure.

   Rough scale of consequence severity (in increasing order):

   - Inconvenience (1–3)
   - Disruption of standard workflow (4)
   - Disruption of critical workflow (5)
   - Complete system failure (6)
   - Complete system failure that spreads to other user systems (7)
   - Property damage (8)
   - Death (9)

3. Multiply the two numbers.

The higher the resulting number, the higher the priority of the test and the more times you need to perform the test to be sure the failure does not happen.

Reporting defects

After you identify a defect during testing, you need to report it to the correct team to repair it. Your report is a type of technical communication and, as such, should follow the writing guidelines, present the information clearly, and (where appropriate) suggest a solution, or at least a possible path to a solution.

At a minimum, defect reports include the following information:

- The name of the test case that identified the defect
- An identifier for the release or build that exhibited the defect
- A cross-reference to the functional or technical specification that defines the feature or function involved
- The exact steps, leaving nothing out and being accurate in all details, that led to discovering the defect (see Chapter 3, *Writing Good Procedures*, for details on how to do this well)
- If it's an *expectation-based defect*—that is, a case where you expected one outcome and got a different one—document what you expected to happen vs. what did happen
- The impact of the defect on the user and/or on the system itself
- Where appropriate, a suggested solution, particularly if it's an expectation-based defect

It can be difficult to tell what is useful diagnostic information. We recommend erring on the side of extreme completeness to be certain you've captured the defect's environment accurately.

Hierarchy of testing documents

Test cases are the specific tests written to test one small part of the product or a group of parts working together. Test cases can be simple or complex, depending on what you're testing and when you're testing it in the development of the product. Earlier testing may be simpler because the product is still simple because less of it is ready. Later tests may be more complex because more of the product is ready to test, and you've tested the small parts individually.

Test cases are rolled into *test suites*. Test suites are groups of related tests. For example, if you're writing test cases for a car, the powertrain is a test suite, and every test about the powertrain is included in that test suite.

Often the test suite is given a whole number, such as Test Suite 4. The test cases in that test suite start with a 4 and are given dot numbers, based on further groupings.

For example, to test a gasket in the powertrain, the test case might be 4.3.2.347. It belongs to Test Suite 4 (powertrain testing) and then other groupings in the test suite. How these are grouped depends on how the company you work for decides that the test suites should be grouped. Everyone working on the testing project knows how these are grouped, and they often refer to the tests and the test suites by numbers alone.

Frequently, the people writing the test cases are not the people who perform the tests. With the international world we have, testing is often done in another office in another time zone, using the value of significant time-zone changes to allow testing to happen while the main office sleeps. Test results appear in the morning email.

Deconstructing test cases

Just as you start by deconstructing features and functions into the smallest sensible specifiable unit, you need to start by deconstructing those specifiable units into the smallest sensible testing unit. Testing units are more granular than specifiable units—any given specifiable unit can, and probably will, generate multiple test cases. It's likely that each response in the stimulus/response table in the specification will require, *at a minimum*, one test case.

Deconstructing spec units into smaller, granular test cases lets you know whether the components work on their own before you can begin testing for whether they work together. We focus on unit testing rather than integration testing (testing multiple components working together); for the purposes of understanding how a test case is constructed, unit test cases suffice.

What makes a good test case?

Test cases are specific, clear, and precise step-by-step instructions for how a person (the tester) performs the test. Anyone should be able to perform the test and get the same results. If you have a background in science, it's like writing an experiment: the experiment must be reproducible to be valid.

Test cases are like recipe directions

Test cases are a lot like written recipes. In theory, if you follow the recipe instructions exactly, your dish is just like the dish someone else makes with that recipe.

Descriptive title: The title lets you know what the dish is in some way. *My Mom's Favorite* is not at all descriptive—you don't even know what sort of dish this is, much less what meal this might be appropriate for. It could be a cocktail or a dessert—you really don't know. When you name your test case, use a descriptive title that makes it clear what you're testing in this test case.

A picture: You need to know what the result should look like. The picture should leave no doubt what the dish looks like. A test case may not have a picture of the result, but it has a clearly defined description of what the result of this test case should be.

List of ingredients and the quantity you need: An outstanding recipe lists the ingredients in the grouping you need them in, such as marinade, sauce, and so on. A really outstanding, excellent recipe tells you the state of each ingredient, such as softened butter. Because the chemical reactions of melted butter are different from softened butter, it matters if you're baking. Not telling you the butter needs to be softened until the **Add the Butter** step is bad. A good test case lists all the equipment you need before you start, such as a ruler. An excellent test case includes the state of that equipment, such as a fabric metric measuring tape.

The equipment you need: An excellent recipe also includes any equipment you might need, such as a stand mixer and an 8" × 8" pan. An excellent test case tells you the state the system must be in before you start. If the test case is an integration test, it tells you what test cases must be completed before you can start this one. A really good test case is specific about what conditions must be met before you can start the test.

List of steps you need to go through: A recipe must list the steps, or procedures, required in the order you need to do them. Good recipes often group the steps into various activities. It tells you the actions you must take as you take them and tells you the measurable or observable result of each step. This lets you know when you're done with that step or set of steps. After reading Chapter 3, *Writing Good Procedures*, this may sound just like a good procedure, and you are correct. Similarly, a good test case includes a list of procedural steps, possibly grouped into activities, with specific observable or measurable results at each step or groups of steps.

Plating or serving instructions: Similarly, in a test case, you get what the final observable or measurable state is at the completion of the test. Unlike a recipe, if you don't get the observable or measurable results described in the test case, the test fails. Unless you completely messed up a recipe, you can usually eat the failed result.

Structure of test cases

The structure of test cases vary from company to company, but they generally all have similar parts. They may look different from company to company as well. While testing professionals have industry best practices, there is a lot of variety. If the way your company creates test cases meets its business needs, then it's the right way. If the number of bugs shipped is higher than industry average for your industry, then the test cases may not be meeting your business needs and should be looked at.

We approach testing features and functions differently in this text because they are different. The distinction we make between feature and function is artificial (see the section titled "The nature of a functional specification" in Chapter 16, *Writing Functional Specifications*), but it is important because how you approach each differs.

Writing test cases for functions and features

Regardless of your actual template, a solid test case includes the elements explored below.

Test plan, suite, case naming and/or numbering

Every test case shows the structure of the overall plan to establish context. Include the test plan name, the test suite name, and the test case name. If you're using hierarchical numbering (which we strongly recommend), include the numbering as well, at the start of the name.

For example, working from the sample spec in Chapter 16, *Writing Functional Specifications*:

Test Plan for Online Banking Mobile App

Test Suite 1001:	Secure Login
Test Case 1001.0001:	Username Field Validation (AJAX)
Test Case 1001.0601:	The login UI is easy to use

Note that the sample Test Case 1001.0001 focuses on testing the AJAX field validation when the user creates a login (such as whether the username is constructed correctly) as opposed to authenticating the username against the security database (in other words, testing whether this is a valid user with an account). Validating login creation and validating existing accounts are *two different test cases*.

Description

As with the specification, the test case description must answer the four user questions, but with a slight adjustment to make it clear that these are testing questions. In many cases, you can provide a simple, single summary sentence.

Notice below, for these component tests, we only need one sentence that follows the clear writing guidelines. One of the ways you know you're writing a component test is you can describe what you're testing with one sentence. When you need to add more sentences, you may be writing an integration test. If you find yourself writing multiple sentences, see whether if you can deconstruct this test further.

For the function test case, you have:

- The Secure Log-In dialog validates Marissa's username against defined field criteria.

For the feature test case, you have:

- The login UI is easy to understand/interact with.

"And" is not your friend here for unit tests. The instant you type the word "and" in this summary, *stop* and rethink the test. You are probably trying to test too many things in one test.

ⓘ The one exception to using "and" in a unit test case is if you provide, as we do in the next example, both the positive and negative outcomes.

What are you testing?

To answer this question, again be literal: what, specifically and concretely, is being tested? For example, in our secure login user dialog function, we can answer the question this way:

- This test ensures that the code correctly validates a new username to verify that it contains only alphanumeric characters and does not contain any special characters.

For the feature test, you can answer the question this way:

- This test verifies that a novice user can locate the fields then log in in under 10 seconds.

Why does [persona] need this to work?

To answer this question, you must review your personas from the spec, identify one (usually the primary persona), and put yourself in her shoes. Why does this person need the function to work correctly? What other features or functions depend on this function working correctly? Continuing our example, we use the following to answer this function question:

- Marissa needs to know immediately after entering her new username that it is valid and can be used to identify her during secure log in. The overall secure login function requires a correctly configured username so that when Marissa logs in to her account, the authentication database can match her username to her account.

We could use the following to answer this feature question:

- Marissa doesn't want to struggle to get logged into her account. She doesn't want a barrier to get to her bank account information.

This is the last place in this test case where we use the persona name. The persona cannot participate in the test because the persona is our made-up invisible friend who does not exist.

How do we test this?

To answer this question, summarize the testing *process*. How are you going to perform the test? Perhaps you test using live testers. Then your *How* should say that and summarize the manual process. You'll put the details in the Action/Result table, discussed below.

Perhaps you test using a script, algorithm, or other automated testing approach. In general, this is a better approach—humans are terrible testing devices. We're not consistent, we're not reliable, we're not at all precise as a rule, and we get bored easily.

If you are planning to test automatically, use this section to describe what the automated test is going to do. Provide enough detail to allow the testing team to build an accurate, effective automated test.

Continuing our function example further, we can use the following to answer this question:

> • To test this function, we set up an automated testing algorithm that enters a predefined set of correct and incorrect usernames into the field and moves to the next field. The algorithm traps the AJAX validation response (correct or incorrect) and prepares a results report.

Note that this test does not verify whether AJAX displays an error message—in a well-deconstructed test suite; that is a separate test. "And" is not your friend when you're designing "smallest sensible unit" testing.

For the feature test, we can say:

> • To test this function, we locate 10 people who have never seen our application before. We give them valid sample login credentials and record how long it takes them to get logged in.

Not all tests are the smallest sensible unit test, though. After you've written all your building block cases, you can start to combine them into integration tests. Integration testing is outside the scope of this book; we don't explore it separately because the basic principles for writing the integration tests are the same as for unit testing.

How do we know this feature or function passed the test?

To answer this question, think about how do you *know* it passed the test? What specific and concrete results do you expect to see to know that the function works to spec? This section must include a measurement number *or* an unmistakably observable outcome. "Works correctly" is neither measurable nor unmistakably observable.

Finishing our example function description, we can use the following to answer this question:

- After running the automated test, we review the results to verify that AJAX correctly identifies properly constructed vs. improperly constructed usernames at least 99% of the time.

Finishing our example feature description, we can use the following to answer this question:

- After running the user test, at least 9 of the 10 users successfully login in under 10 seconds.

Materials

In this section, provide a complete and thorough list of the materials your tester needs to perform the test, including any tools needed to measure time or distance. In some testing labs, some test set-ups are so common they are given names, like Float Test Station 1.

In our function example, we list the following:

- Predefined list of 50 usernames: 25 correctly configured, 25 incorrectly configured
- Automated testing algorithm for AJAX field-testing, configured with the username field requirements specified in [*name of spec document*]
- Testing computer capable of running the testing algorithm

In our feature example, we list the following:

- 10 product-naive volunteers
- Predefined list of 10 correctly configured usernames and passwords
- Each username and login information recorded on a separate sheet of paper
- Usability testing lab set up for 10 users in isolation
- Computers with the user-ready UI for correctly logging in
- Timers

Because we're usability testing and because we specified the usability testing lab, the lab has cameras that record the volunteer actions and faces as well as software that records what they type on the screen.

Prerequisite tests

In this section, list any tests that the product must have passed before we can effectively test the function in question.

Our functional example is a foundational test; that is, there are no prior tests required. If we were instead testing that the **Create Account** button changes from inactive to active from our sample function, the prerequisites would include our example test and the password field validation.

Our feature test is a usability test of the Login screen and the Logged-in screen. However, we can't do this test until we have verified the login tests work. We could, in the interest of time, create mock-ups of these screens. If that's the case, we list those in the Materials-needed section.

Many tests require other parts of the product to work before we can run this test. For example, for a physical device, we first need to test:

- whether power is coming from the **On** switch because we must be sure that device can be turned on, *and*
- whether the device can boot or run through the start-up sequence, *and*
- whether the home screen appears on the panel so we know that after the start-up sequence occurs, it completes properly.

For a software product, we first must test:

- whether the Welcome screen appears, *and*
- whether an account can be created, *and*
- whether that account can be used to log in

before we can test any features that require a user to be logged in.

All these tests must be run and passed before we can isolate our variables enough to confidently run other tests.

Action/result tables

The action/result table details the steps the tester (not the persona, who is fictional and can't test anything) must take to perform the test. Because you're creating direct instructions to the tester, use the imperative voice in the Action column. For more information about the imperative voice, see Chapter 2, *Clear Writing Guidelines*.

If you are designing a *live-person manual* test, this table describes each step the tester must do to complete the test.

If you are designing an *automated* test, this table describes each step the tester must do to run the automation.

> ⓘ This tester is not the persona name because the persona is our made-up invisible friend who does not exist. The persona cannot run any tests. The testing staff follow the steps in this section exactly as you describe them.

Recall Chapter 3, *Writing Good Procedures*. The same structure of numbered steps applies here as well, but pulled apart for clarity. In this structure, put the exact action the tester takes on the left side of the table. Put the observable or measurable result of that action on the right side.

Note that the action/result relationship in the testing procedure is one-to-one. Testers need to know at each step what the expected observable or measurable result is. If that observable or measurable result is not exactly the result the tester gets, the test is over, regardless of where the tester is in the testing steps.

The following example shows an action/result table for our sample function.

Action	Result
Start the testing program.	The interface shows a prompt to specify the field name to be checked. [*include screen capture or mockup of the prompt*]
At the prompt, type *Username*.	The interface shows a prompt to enter the username file to be uploaded. [*include screen capture or mockup of the prompt*]
At the prompt, enter the name of the test data file, UsernameTest.txt, which contains 100 usernames, then click **Enter**.	The interface prompts to enter the number of times to run the test. [*include screen capture or mockup of the prompt*]
At the prompt, type *1* (because the test data file has 100 names, so, in effect, you're running the test 100 times), then click **Enter**.	The interface begins running and shows a progress bar. When complete, it presents a results report. [*include screen capture or mockup of the prompt*]
Review the results.	For at least 99 out of 100 user names, the report shows that the results are **True** for each correct username and **False** for each incorrect username.

The following example shows an action/result table for our sample feature.

Action	Result
Identify the login information on a sheet of paper for the volunteer. Ask them to use that information to log in. Tell them to start.	Start the timer.
The volunteer logs in.	Stop the timer.
Record the time results.	The time for each volunteer appears on the recording sheet.
Review the results from all volunteers.	At least 9 of the 10 successfully logged in in under 10 seconds.

Target pass/fail metric

Restate the pass/fail conditions from the description, expressed as a ratio: 99/100. Do not express as a percentage, which is not specific enough. Exactly how many times are you running the test (the denominator)? How many times must the test succeed before you can say with confidence that the feature or function performs to spec (the numerator)?

> **A note about automated software testing:** With automated testing, you typically run the test one time, with the number of records in the test file representing the effective number of tests. For example, if you're testing login, as in the function example above, you have 100 names in the test file to run the test 100 times. Your denominator reflects the number of *tests* (100), not the number of times you *ran the program* (1).

Review the feature or function's assigned priority, then determine the risk associated with the function (see the section titled "Risk analysis and its effect on testing" (p. 230)). The higher the priority, combined with the higher the risk, the more test iterations you should plan to run.

> **A note about perfection:** You won't get it, because you're human. Do not set up a ratio of 100/100, 1000/1000, and so on. If you need more precision, run the test more often, and shoot for "five nines" (99,999/100,000). Even medical devices and airplanes specify five nines in their testing. Perfection is not obtainable nor is it a goal in testing.

Metaphors

*Analogies illuminate complicated topics by comparing them to objects we understand, experiences
we have had, or situations we can imagine.*
—Anne Janzer, Writing to Be Understood (Janzer 2018, page 107)

Humans love stories. It seems to be a cross-cultural thing, in that all humans in all cultures love stories. We're hard-wired for a good story. That also means we're hard-wired for imagery. And the best imagery includes metaphors.

What is a metaphor?

Metaphors are "a figure of speech in which a word or phrase literally denoting one kind of object or idea is used in place of another to suggest a likeness or analogy between them."[1] You may have been taught that a metaphor is different from a simile which is different from an analogy, and they are. We include all three in the metaphor category because, for our purposes, it is a distinction that makes no difference. All three are what is called "figurative language."

All language is full of metaphors. Every time we say "this is like that," we use a metaphor. For example: "It is the east, and Juliet is the sun"[2] is a metaphor. Shakespeare did not mean that Juliet is a large gaseous thermonuclear ball that hangs in space, slowly coming over the horizon as you stand on a planet that orbits that sun. This is a metaphor, in that Juliet is warmth and light and the source of life, the center of all good things. She is like a sunrise when the sun's rays are visible although the sun itself is not yet visible. Her very arrival brings light to darkness.

Metaphors are language and culture specific. You know this if you've ever seen something translated literally from another language. For example, most swearing in other languages doesn't make a lot of sense when translated exactly to English. Swearing is often a metaphorical description, closely tied to the culture of the language. For example, the Russian swear word *Развалюха* (Razvaluha) translates literally to "car that's falling apart as it goes."[3] We can see the metaphor, even though we may not speak Russian, but the full meaning doesn't come through.

---

[1] https://www.merriam-webster.com/dictionary/metaphor

[2] Romeo & Juliet, Act 2, Scene 2, Lines 2–6

[3] https://www.thetraveltart.com/russian-swear-words-slang-expletives/

But metaphors aren't just for swearing or stories—they can help us better understand a concept or frame a complex situation, especially in technical language. Understanding how and when to use metaphors in your work is important because it can help make concepts clearer and more relevant to the reader.

When you use a relevant metaphor, the reader's brain reacts to that metaphor as though that thing is happening to them. For example, if you put someone in an MRI and tell them the phrase "He's so sweet," the part of their brain associated with taste lights up.[4] Our brains act it out.

Not all metaphors are created equal

Both sides of the brain are involved in understanding metaphors and figurative language. The specific areas of the brain involved depend on:

- **Figurativeness:** how literal the metaphor is,
- **Familiarity:** does the reader have knowledge about this, and
- **Difficulty:** how hard is it to understand the metaphor.[5]

Unique metaphors—ones that put words together in a new (novel) way—involve more of the brain than more common metaphors.

What this means for us is that metaphors help our readers understand what we mean. And that if we use metaphors that require less cognitive processing, they understand more easily. Familiar metaphors (but not old and worn out metaphors) are easier to understand than unfamiliar metaphors, as the familiar ones use less cognitive processing power.

When we create technical information, we often need a metaphor to explain a concept. Unlike literature, where we can expect our readers to be fully cognitively engulfed in the reading experience, our audience is typically under cognitive load. This research tells us that more familiar metaphors are easier to understand.

---

[4] "Concrete processing of action metaphors" (Lai 2019)

[5] "Neural correlates of metaphor processing" (Schmidt 2009)

Personas drive metaphors

When you communicate you need to use metaphors, especially metaphors chosen to resonate with your readers. How do you know what the right metaphors are? Your personas drive this.

Sharon worked on railway crossing technology at one point in her career. The company hadn't defined personas. She asked the training group if she could do a survey to get her arms around who the users were. Sometimes surveys are a great place to start finding out who your users are.

In a moment of inspiration, she included several questions about the users' employment history. She asked what job roles they had before they became railway maintenance staff. As she got the results back, she discovered these men (and they were all men) were in the trades before they started working in train-crossing maintenance. The domains of experience these men had were firmly in construction, electrical, and mechanics.

This insight allowed Sharon to create content that used metaphors and examples directly from the trades to help the instructions make more sense to the users. These metaphors and examples were easier for the maintenance workers to understand because they drew from the worker's domain knowledge.

These same metaphors and examples would not have worked with non-tradespeople because they lack the domain experience. For example, a person who doesn't know how to rewire a circuit might not understand these metaphors. That person doesn't have the domain experience to know these metaphors or examples.

Metaphors and examples

Examples may resonate in the brain in a way that's similar to metaphors. We suspect that when you use an example that resonates with the reader, the brain may light up in a way that is similar to metaphors. We know of no research done on what our brains do when faced with examples (if you do, please contact us!). But it makes inherent sense to us that examples work the same areas of our brains that metaphors engage.

Why? Because examples are about concrete things the reader knows about but presented in a metaphorical way. An example tells a story that illustrates an idea and makes that idea more concrete. Readers don't, at that moment, experience the story described in the example, but they may be familiar with the example. To get very meta, let's look at the example used above.

In the story about the train-crossing maintenance workers, you read about things you are familiar with, even if you don't know anything about train-crossing equipment. You may never have created a survey, but you've probably taken one in your lifetime. You never met the men she wrote about and may never have worked in the trades, but you may have met people like them. And ultimately, the story puts you in Sharon's life and experience and helps explain a concept.

You can easily imagine all these things. Sharon used this story to explain how personas connect to technical content and metaphors. That's metaphorical.

APPENDIX B

References

The field of communication and engineering is rich and deep. This section is not a complete list of all resources available, but rather some exceptional sources that can help you learn more.

Chapter 1: *Welcome*

Hensel, Danny. 2020. "Are Your Texts Passive-Aggressive?: The Answer May Lie In Your Punctuation." https://www.npr.org/2020/09/05/909969004/-before-texting-your-kid-make-sure-to-double-check-your-punctuation

NASA. 1999. "Mars Climate Orbiter Mishap Investigation Board Phase I Report." PDF format. https://-llis.nasa.gov/llis_lib/pdf/1009464main1_0641-mr.pdf

Chapter 2: *Clear Writing Guidelines*

Janzer, Anne. 2018. *Writing to Be Understood: What Works and Why*. Cuesta Park Consulting.

New Scientist. 2016. *How Long is Now: Fascinating answers to 191 Mind-boggling questions*. Nicholas Brealey Publishing.

Moran, Kate. 2016. "How Chunking Helps Content Processing." https://www.nngroup.com/articles/chunking/

Pernice, Kara. 2019. "The Layer-Cake Pattern of Scanning Content on the Web." https://www.nngroup.com/-articles/layer-cake-pattern-scanning/

Purdue OWL Labs. "Active Versus Passive Voice." https://owl.purdue.edu/owl/general_writing/-academic_writing/active_and_passive_voice/active_versus_passive_voice.html

Schade, Amy. 2018. "Inverted Pyramid: Writing for Comprehension." https://www.nngroup.com/articles/-inverted-pyramid/

Schriver, Karen, Annetta L. Cheek, and Melodee Mercer. "The research basis of plain language techniques: Implications for establishing standards." *Clarity, Journal of the international association for promoting plain legal language* 63 (May 02, 2010): 26-32. https://www.researchgate.net/publication/-285927928_The_research_basis_of_plain_language_techniques_Implications_for_establishing_standards

Swisher, Val. 2015. "Three Reasons to Optimize Your Content." https://contentrules.com/-three-reasons-optimize-content/

U.S. Government. "Federal plain language guidelines." https://www.plainlanguage.gov/guidelines/

Vincent, Sara. 2014. "Sentence length: why 25 words is our limit." https://insidegovuk.blog.gov.uk/2014/08/-04/sentence-length-why-25-words-is-our-limit/

Wylie, Ann. 2020. "How long should a sentence be?" https://freewritingtips.wyliecomm.com/2020-03-19/

Chapter 3: *Writing Good Procedures*

Godin, Seth. 2023. "Thoughts on the manual." https://seths.blog/2023/09/thoughts-on-the-manual/

Chapter 4: *Writing Tools*

Gray, Marks. 2021. "Tips On Consistency For Trademark Owners." https://marksgray.com/-intellectual-property/trademark-consistency/

Harvard University. 2023. "Getting started with prompts for text-based Generative AI tools." https://-www.huit.harvard.edu/news/ai-prompts

IBM. 2023. "What are AI hallucinations?" https://www.ibm.com/think/topics/ai-hallucinations

Patti, Keaton. 2020. *I Forced a Bot to Write This Book: A.I. Meets B.S.* Andrews McMeel Publishing.

Schneier, Bruce, and Nathan E. Sanders. 2025. "AI Mistakes Are Very Different From Human Mistakes." https://spectrum.ieee.org/ai-mistakes-schneier

Timofejeva, Gita. 2024. "Translation costs: rates per word and how to lower them [+ free calculator]." https://lokalise.com/blog/lower-localization-costs/

Chapter 5: *The Business Context of Communication*

Burke, Dan. 2002. *Business @ the Speed of Stupid.* Basic Books.

Charon, Ram. 2017. *What Your CEO Wants You to Know.* Currency.

Labmate. "What is 'Blue Sky Science'?" https://www.labmate-online.com/news/news-and-views/5/-breaking-news/what-is-lsquoblue-sky-sciencersquo/30187

Marchand, Lisa. 2017. "What is readability and why should content editors care about it?" https://-centerforplainlanguage.org/what-is-readability/

Moore, Geoffrey. 2005. *Inside the Tornado: Strategies for Developing, Leveraging, and Surviving Hypergrowth Markets.* 3rd ed. Harper Business.

Moore, Geoffrey. 2014. *Crossing the Chasm*. 3rd ed. Harper Business.

Pernice, Kara. 2017. "F-Shaped Pattern of Reading on the Web: Misunderstood, But Still Relevant (Even on Mobile)." https://www.nngroup.com/articles/f-shaped-pattern-reading-web-content/

Rogers, Everett M. 2003. *Diffusion of Innovations*. 5th ed. Free Press. The breakdown of adopters by percentage is shown in a graph on page 181 of this edition.

Wikipedia. 2022. "Disruptive innovation." https://en.wikipedia.org/wiki/Disruptive_innovation

Chapter 6: *The Workplace Ecosystem*

Annie E. Casey Foundation. 2023. "What's the Difference Between Equity and Equality?" https://www.aecf.org/blog/equity-vs-equality

Doran, George T. "There's a S.M.A.R.T. Way to Write Management's Goals and Objectives." *Management Review* 70, no. 11 (1981): 35–36.

Favis, Elise, and Mikhail Klimentov. 2021. "Cyberpunk 2077's launch, explained: CD Projekt Red releases apology video." Behind The Washington Post's paywall. https://www.washingtonpost.com/video-games/2020/12/15/controversies-cyberpunk-2077s-bungled-launch-explained/

Lencioni, Patrick, and Kensuke Okabayashi. 2008. *The Five Dysfunctions of a Team: An Illustrated Leadership Fable*. John Wiley & Sons (Asia).

Mindful Staff. 2024. "Zoom Fatigue is Real: Here Are Six Ways to Find Balance." https://www.mindful.org/zoom-fatigue-is-real-here-are-six-ways-to-find-balance/

Mindtools. "The COIN Conversation Model." https://www.mindtools.com/a94k5vp/the-coin-conversation-model

Chapter 7: *Résumés and Cover Letters*

Turczynski, Bart. 2022. "2022 HR Statistics: Job Search, Hiring, Recruiting & Interviews." This reference contains a wide variety of statistics from multiple sources. https://zety.com/blog/hr-statistics

Chapter 8: *Ethics in Engineering*

El-Zein, Abbas. 2013. "As engineers, we must consider the ethical implications of our work." https://www.theguardian.com/commentisfree/2013/dec/05/engineering-moral-effects-technology-impact

Gladwell, Malcolm. 2019. *Talking to Strangers: What We Should Know about the People We Don't Know*. Little, Brown and Company.

Chapter 10: *Pitching Ideas*

Carnevale, Anthony P., Stephen J. Rose, and Ban Cheah. 2011. "The College Payoff: Education, Occupations, Lifetime Earnings." Includes link to full report PDF. https://cew.georgetown.edu/cew-reports/-the-college-payoff/

Luthi, Ben. 2025. "The Average Car Price Is Nearing All-Time High." https://www.experian.com/blogs/-ask-experian/average-car-price/

Chapter 11: *Designing Effective Presentations*

ARTnews. 2022. "The Best Color Theory Books for Foundational Knowledge." https://www.artnews.com/-art-news/product-recommendations/best-color-theory-books-1202694928/

Burton, Sharon. 2012. *8 Steps to Amazing Webinars*. XML Press.

Chappell, Bill. 2015. "NFL's Red And Green Uniforms Described As 'Torture' By Colorblind Fans." https://www.npr.org/sections/thetwo-way/2015/11/13/455896618/-nfls-red-and-green-uniforms-described-as-torture-by-colorblind-fans

Duarte, Nancy. 2018. *Slide:ology: The art and science of creating great presentations*. O'Reilly Media.

Horton, William. 1991. *Illustrating Computer Documentation*. Wiley and Sons.

Lile, Samantha. 2020. "44 Types of Graphs Perfect for Every Top Industry." https://visme.co/blog/-types-of-graphs/

Mandal, Ananya. 2019. "Color Blindness Prevalence." https://www.news-medical.net/health/-Color-Blindness-Prevalence.aspx

Redish, Janice C. (Ginny). "What Is Information Design?" *Technical Communication*, 2000, no. Spring 2000:163–166.

Statistic Brain Research Institute. 2016. "Fear of Public Speaking Statistics." Subscription required. https://www.statisticbrain.com/fear-of-public-speaking-statistics/

Tufte, Edward. 1997. *Visual Explanations*. Graphics Press.

Tufte, Edward. 2001. *The Visual Display of Quantitative Information*. 2nd ed. Graphics Press.

WebAIM. 2016. "Contrast Checker." Website that checks foreground and background colors for contrast. Shows how well the colors score against WCAG AA and WCAG AAA (Web Content Accessibility Guidelines) levels suggested by the WorldWide Web Consortium (https://www.w3.org/). https://www.statisticbrain.com/fear-of-public-speaking-statistics/

Chapter 12: *Handling Yourself and the Room in Presentations*

Dvorsky, George. 2012. "The neuroscience of stage fright — and how to cope with it." https://gizmodo.com/-the-neuroscience-of-stage-fright-and-how-to-cope-with-5950544

Chapter 13: *Cognitive Science*

Berlin, Brent, and Paul Kay. 1969. *Basic Color Terms: Their Universality and Evolution*. University of California Press.

Goldstein, E. Bruce, and James R. Brockmole. 2017. *Sensation and Perception*. Cengage Learning.

Kister, Tina. 2019. "Perception and Design: A Science-Based Approach for Creating Content that Works." https://www.brighttalk.com/webcast/9273/364139

Chapter 14: *Constructing Explanations*

BBC Earth. 2014. "Are Crows the Ultimate Problem Solvers?" YouTube video. https://www.youtube.com/-watch?v=cbSu2PXOTOc

Bradley, Steven. 2014. "Design Principles: Visual Perception And The Principles Of Gestalt." Part of a series on design principles. https://www.smashingmagazine.com/2014/03/-design-principles-visual-perception-and-the-principles-of-gestalt/

Dolan, Eric W. 2017. "Study reveals just how quickly we form a first impression." https://www.psypost.org/-2017/10/study-reveals-just-quickly-form-first-impression-50039

Firtina, Nergis. 2022. "Apes can make rational, economic decisions as humans do, research finds." https://-interestingengineering.com/science/ape-rational-economic-decisions

Heckel, Paul. 1991. *The Elements of Friendly Software Design*. SYBEX.

Laurel, Brenda, and S. Joy Mountford, eds. 1990. *The Art of Human-computer Interface Design*. Addison-Wesley.

Lim, Jonathan B., and Daniel M. Oppenheimer. 2020. "Explanatory preferences for complexity matching." https://www.ncbi.nlm.nih.gov/pmc/articles/PMC7173929/

National Geographic. 2022. "Does Tool Use Define Humanity?" Video format. https://-education.nationalgeographic.org/resource/do-tools-make-man/

Norman, Don. 2013. *The Design of Everyday Things*. Revised and expanded edition. Basic Books.

Seeker. 2015. "Which Animals Recognize Themselves In Mirrors?" YouTube video. https://www.youtube.com/-watch?v=cKs_iW0QVNY

Sellen, Abigail, and Anne Nicol. "Building User-centered On-line Help." In *The Art of Human-Computer Interface Design*, edited by Brenda Laurel, 145. Addison-Wesley Publishing, 1990.

Tabarrok, Alex. 2018. "Dolphin Capital Theory." https://marginalrevolution.com/marginalrevolution/2018/-01/dolphin-capital-theory.html

Tyson, Neil deGrasse. 2017. "Neil deGrasse Tyson explains how aliens could be so much smarter than us." Video format (YouTube). https://www.youtube.com/watch?v=Mvfi49XM5_s

Tyson, Neil deGrasse. 2020. "Neil deGrasse Tyson Explains Why Some Info Is Need to Know." Video format (YouTube). https://www.youtube.com/watch?v=RQj0cxHJ4fc

Chapter 15: *Personas and Scenarios*

Bell, Edward. 2021. LinkedIn. Edward Bell's LinkedIn page. https://www.linkedin.com/posts/-edward-bell-86309024_if-you-were-segmenting-based-on-demographics-activity-6869489639191928832-7Wcw/

Cooper, Alan. 2004. *The Inmates are Running the Asylum: Why High Tech Products Drive Us Crazy and How to Restore the Sanity*. Sams - Pearson Education.

Furey, William. "The Stubborn Myth of 'Learning Styles': State teacher-license prep materials peddle a debunked theory." *Education Next* 20, no. 3 (2020): 8-12. https://www.educationnext.org/-stubborn-myth-learning-styles-state-teacher-license-prep-materials-debunked-theory/

Langreo, Lauraine. 2024. "U.S. Students' Computer Literacy Performance Drops." https://www.edweek.org/-technology/u-s-students-computer-literacy-performance-drops/2024/12

Nielsen, Jakob. 2016. "The Distribution of Users' Computer Skills: Worse Than You Think." https://-www.nngroup.com/articles/computer-skill-levels/

Sarabi-Asiabar, Ali, Mehdi Jafari, Jamil Sadeghifar, Shahram Tofighi, Rouhollah Zaboli, Hadi Peyman, Mohammad Salimi, and Lida Shams. "The Relationship Between Learning Style Preferences and Gender, Educational Major and Status in First Year Medical Students: A Survey Study From Iran." *Iranian Red Crescent Medical Journal* 17, no. 1 (December 27, 2014). https://pmc.ncbi.nlm.nih.gov/-articles/PMC4341501/

VARK Learn Limited. 2021. "VARK: helping you learn better." Website about the VARK learning styles questionnaire. https://vark-learn.com

Zerrenner, Emily. 2024. "Turn it off and on again: digital literacy in college students." https://acrlog.org/-2024/03/21/turn-it-off-and-on-again-digital-literacy-in-college-students/

Chapter 16: *Writing Functional Specifications*

Davis, Alan M. 1993. *Software Requirements: Objects, Functions, and States*. Facsimile, Subsequent edition. Prentice Hall.

Gause, Donald C., and Gerald M. Weinberg. 2011. *Exploring Requirements 1: Quality Before Design*. Weinberg & Weinberg.

Gilb, Tom. 1988. *Principles of Software Engineering Management*. Addison-Wesley Professional.

Gilb, Tom. 2005. *Competitive Engineering: A Handbook for Systems and Software Engineering Management*. Butterworth-Heinemann.

Jackson, Michael. 2000. *Problem Frames: Analyzing and Structuring Software Development Problems*. Addison-Wesley.

Kovitz, Benjamin L. 1998. *Practical Software Requirements: A Manual of Content and Style*. Manning Publications.

Wiegers, Karl E. 1999. *IEEE Software Requirements Template*. IEEE. Downloadable Microsoft Word file. https://web.cs.dal.ca/~hawkey/3130/srs_template-ieee.doc

Chapter 17: *Testing Your Products*

RTI. 2002. "The Economic Impacts of Inadequate Infrastructure for Software Testing." PDF file (over 300 pages). https://www.nist.gov/document/report02-3pdf

Appendix A: *Metaphors*

Janzer, Anne. 2018. *Writing to Be Understood: What Works and Why*. Cuesta Park Consulting.

Lai, Vicky T., Olivia Howerton, and Rutvik H. Desai. "Concrete processing of action metaphors: Evidence from ERP." *Brain Research* 1714 (July 01, 2019): 202–209. https://www.sciencedirect.com/science/-article/abs/pii/S0006899319301283

Schmidt, Gwenda L., and Carol A. Seger. "Neural correlates of metaphor processing: the roles of figurativeness, familiarity and difficulty." *Brain and Cognition* 71, no. 3 (December 2009): 375–386. https://-pmc.ncbi.nlm.nih.gov/articles/PMC2783884/

Index

Symbols

4 user questions, 192
    designing and, 196
    examples: functional spec, 197
    examples: test case, 199
    examples: user assistance, 196
    how do I do it, 194
    how do we design it, 221
    how do we know it passed, 239
    how do we test it, 238
    procedures and, 194
    what can go wrong, 221
    what is it, 193, 219, 237
    why did it do that, 195
    why does persona need this to work, 237
    why does persona want, 220
    why should I care, 192

A

academic ecosystem, 69
academic writing, 7, 13, 57–58
accessibility
    color, 153, 155
    contrast, 155
action/result tables, 195, 243
    functional specifications and, 223
    test cases and and, 241
active perception, 188
active voice, 10–11
    and action verbs, 10
address, business letter, 90

adoption, technology, 132, 136
    schemas and, 179
agile environments, 4
    estimating projects in, 123
    functional specifications in, 117, 215
    project flow in, 113
    project roles in, 114
    scenarios in, 209, 211
    testing in, 120, 227–228, 230
    user stories in, 209, 211
AI, 51
    cautions, 56
    ethical use of, 55
    prompts, 53
    use cases, 55
alert color, 154
algorithms, test cases and, 238, 241
all caps text, 152
analogies, 245
    difficulty, 246
    familiarity, 246
    figurativeness, 246
    personas and, 247
    vs. examples, 247
analyzing risk, 230
animation, 149–150
applications, demonstrating, 164
applying clear writing, 27
    functional specifications, 217
    technical specifications, 217
    user assistance, 216
architect
    development/testing, 121

product managers, 121
project managers, 121
specifications, 118
artificial intelligence (*see* AI)
attentive processes, 174
    cognitive load, 174
attributes/emphasis for slides/UI, font, 151–152
    all caps, 152
    bold, 151
    combination, 151
    italics, 151
    underlining, 151
audience, 7–8, 57, 201
    (*see also* personas or scenarios)
    academic, 58
    addressing your, 12–13
    appropriate reading level for your, 14
    business, 7
    interacting with your, 166–169
        empty stage space, 168
        hand gestures, 167
        moving around, 167
        posture, 167
        reading slides, 168
        verbal tics, 168
        voice, 167
        whole room, 166
        Z sweep, 166
    managers as the, 58
    peers/professors as the, 57
    presenting to an, 145–160
    researching, 207
    scripted presentations, 162
    testing presentations with an, 162
    using color to influence your, 189
    using schemas to connect with your, 179
aural/auditory input mode, VARK and, 202–203
automated test cases, 238, 241
automatic behavior, 173
    habits and, 180
    standards and, 175
awards in résumés, 98

B

Bailie, Rahel, viii
bar charts, 158
behavior of humans, 171–181
    attentive processes, 174
    brains and, 173
    cocktail party effect, 176
    complex vs. simple explanations, 183
    explanations, 183–184
    learning, 177
        cognitive load, 180
        experience, 178
        habits, 180
        interference, 180
        schemas, 178
    physical world and, 171
    preattentive processes, 173
        habits and, 180
        standards and, 175
    redundant signals and, 177
    sensory adaptation, 177
    systems of information and and, 171–172
    vision and, 173, 185
        black space on page, 186
        gray space on page, 186
        space on page, 186
        white space on page, 186
biology, human behavior and, 171–172
    attentive processes, 174
    brains, 173
    cocktail party effect, 176
    explanations, 183–184
    preattentive processes, 173
        standards and, 175
    redundant signals and, 177
    sensory adaptation, 177
    vision, 173, 185
        gray space on page, 186
        space on page, 186
black space, 186
    in written documents, 187

bold text, 151
bosses, writing for, 58
brain rest, 186
brains, human, 173
    attentive processes, 174
    cocktail party effect, 176
    cognitive explanations, 191
        designing with user questions, 196
        examples: functional spec, 197
        examples: test case, 199
        examples: user assistance, 196
        how do I do it and, 194
        procedures and, 194
        user questions and, 192
        what is it and, 193
        why did it do that and, 195
        why should I care and, 192
    complex vs. simple explanations and, 183
    explanations and, 183–184
    learning and, 177
        cognitive load, 180
        experience, 178
        habits, 180
        interference, 180
        schemas, 178
    metaphors/analogies/similes and, 246
    metaphors/analogies/similes vs. examples, 247
    need to know why and, 183
    perceptual explanations, 185
    preattentive processes, 173
        standards and, 175
    redundant signals and, 177
    sensory adaptation, 177
    vs. animals, 184
bubble help, 193
build master, delivery, 125
bullet points on slides, 149
business cases, 137
    cost-avoidance-based, 140–143
        business example, 140
        personal example, 140
    examples

        business, 139–140
        email, 142
        personal, 140
    pitching ideas and, 137
        finance context, 129
        historical/technical context, 131
    résumés and, 89
    ROI-based, 138–139
        business example, 139
        personal example, 138
    technology adoption curve and, 136
    value and, 137
business context, 58
    pitching ideas, 129
business letters, 90
    structure of, 90
        address, 90
        body of letter, 91
        close of letter, 91
        date, 91
        salutation, 91
business requirements, 116–117
    functional specification and, 224
business writing, 7, 10, 13–14
    applying, 27
        functional specifications, 217
        technical specifications, 217
        user assistance, 216
    context of, 57
    noise vs. signal, 15
    procedures, 27–48
    profit and, 58
business writing vs. academic writing, 7, 57–58

C

calculating risk, 231
CEO
    business requirements, 116
    project start, 114
CFO
    business requirements, 116

project start, 115
charts, 157–159
    bar, 158
    formatting, 159
    grammar and spelling in, 160
    line graphs, 158
    pie, 159
chasm, crossing the, 134
chat, 81
    group, 82
    presentations and, 164
    privacy and, 82
    questions
        handling in presentations, 168
Christensen, Clayton, 133
chunked procedures, 43
chunking text, 20
close of business letter, 91
clutter, 146
CMO
    business requirements, 116
    project start, 115
cocktail party effect, 176
cognitive explanations, 191
    user questions and, 192
        designing with, 196
        examples: functional spec, 197
        examples: test case, 199
        examples: user assistance, 196
        how do I do it, 194
        procedures and, 194
        what is it, 193
        why did it do that, 195
        why should I care, 192
cognitive load, 15, 246
    colors and, 153
    graphs/charts and, 157
    impact of input modes on, 205
    learning theory and, 180
    metaphors, 246
    schemas and, 179
    short paragraphs and, 18

    short sections and, 20
    short sentences and, 16
    using standards to reduce, 175
    using white space to limit, 186
cognitive science, 171–181
    attentive processes, 174
    black space on page and, 186
    brains and, 173
    cocktail party effect, 176
    cognitive explanations, 191
    cognitive load, 180
    complex vs. simple explanations, 183
    designing with user questions, 196
    experience, 178
    explanations, 183–184
    gray space on page and, 186
    habits, 180
    how do I do it, 194
    interference, 180
    learning, 177
    metaphors/analogies/similes, 246
    metaphors/analogies/similes vs. examples, 247
    need to know why, 183
    perceptual explanations, 185
    physical world and, 171
    preattentive processes, 173
    procedures and, 194
    redundant signals and, 177
    schemas, 178
    sensory adaptation, 177
    space on page and, 186
    standards and, 175
    systems of information and, 171–172
    user questions, 192
    vision and, 173, 185
    vs. animals, 184
    what is it, 193
    white space on page and, 186
    why did it do that, 195
    why should I care, 192
COGS (see Cost of Goods Sold (COGS))
COIN method, 86

leadership and, 86
color
    accessibility, 153, 155
    as a redundant signal, 153
    blindness, 153
    complementary, 155
    cultural considerations, 173
    dominant, 156
    grouping by, 190
    human perception of, 173
    in slide/UI design, 153–157
        contrast, 155
        mood, 156
        structure/emphasis, 156
    influencing emotion with, 189
    perception of, 154
    recessive, 156
    red = danger, 154
    tone, 155
color rush, 154
color wheel, 155
combining overview and task-specific procedures, 41
common errors in procedures, 37
communication
    aural/auditory input mode, 203
    financial, 131
    guidelines for, 7–25
    kinesthetic input mode, 204
    reading input mode, 204
    teams and, 85
    visual input mode, 203
complementary colors, 155
complex explanations, 183
complex procedures, 42
concise writing, importance of, 15
conclusion slide, 148
conditioned responses, 180
    interference and, 180
consciousness
    attentive processes and, 174
    cognitive load and, 174
    preattentive processes and, 174

conservative adopters, 135
consistency, 150
    in grouping/organizing, 191
    in writing, 13
constraints
    personas and, 208, 213
    scenarios and, 213
contact information, business letter, 90
context, business, 58
    pitching ideas, 129
contrast, color
    accessibility and, 153
    in slide/UI design, 155
contributor, 69
corporate politics during Q&A sessions, 169
cost avoidance, 140–142
    business example, 140
    personal example, 140
Cost of Goods Sold (COGS), 62–63
    graphic representation of, 159
    role in calculating ROI, 138–139
costs
    fixed, 64–65
    G&A, 64–65
    overhead, 64–65
    pro-rated, 64
    variable, 62
cover letters, 90–92
creating
    functional specifications, 215
        deconstructing functions/features, 217
        structure of, 216–217
    meaning
        emotions and, 188
        figure and ground, 189
        grouping and, 190
        perception and, 188–189
    personas, 207–208, 213
    scenarios, 213
    test cases
        deconstructing functions/features, 233
        structure of, 235

testing documents, 227
crunch time, 124
CTO
    business requirements, 116
    project start, 115
culture
    human behavior and, 171–172
    metaphors/analogies/similes and, 245
currency, 59

D

da Vinci, Leonardo, 131
danger color, 154
date, business letter, 91
day of presentations, 164–165
    finishing up, 170
deadlines, 77
deconstructing
    functions/features, 217
    test cases, 233
defects
    costs of, 229
    definition of, 228
    reporting, 232
definitions
    attentive processes, 174
    black space, 186
    cocktail party effect, 176
    experience, 178
    figure, 189
    gray space, 186
    ground, 189
    grouping, 190
    habits, 180
    interference, 180
    page, 185
    preattentive processes, 173
    schemas, 178
    sensory adaptation, 177
    white space, 186
deliberate behavior, 174

delivery of presentations, 161–170
    day of, 164
    script reading, 162
    test run, 162
delivery phase, 124
demographics and personas, 207
description
    functional specifications and, 219
    test cases and, 236
designing
    black space on page, 186
    gray space on page, 186
    grouping and, 191
    perception and, 188–189
    visual space on page, 186
    white space on page, 186
designing interfaces, 145
    animation and, 150
    color and, 153
        contrast, 155
        mood, 156
        structure/emphasis, 156
    consistency and, 150
    font attributes/emphasis and, 151
    font size and, 151
    graphs/charts and, 157
    maximum information and, 149–150
    progressive disclosure and, 149
    serif vs. sans serif, 152
    short text, 149
    text formatting and, 151
designing slides, 145
    accessibility and, 153
    clutter, 146
    colors, 153
        contrast, 155
        mood, 156
        structure/emphasis, 156
    flow, 146
    graphs/charts and, 157
    projected/emitted light, 147
    reflected light, 147

simplicity, 146
structure, 148
text formatting, 151
designing UI/UX, 145
  animation and, 150
  color and, 153
    contrast, 155
    mood, 156
    structure/emphasis, 156
  consistency and, 150
  font attributes/emphasis and, 151
  font size and, 151
  graphs/charts and, 157
  maximum information and, 149–150
  progressive disclosure and, 149
  serif vs. sans serif, 152
  short text, 149
  text formatting and, 151
developers
  delivery, 125
  development/testing, 121
  downtime, 141
  specifications, 118
development
  for "everyone", 201
  products
    personal preferences and, 201–202
    user-centered, 201, 213
      personas, 206–208
      scenarios, 209–212
      wrong personas/scenarios, 212
    using personas in product, 208, 213
    using scenarios in product, 213
development phase, 120, 124
disruptive innovation, 133
disruptive technology, schemas and, 179
document to the question, 191
  designing and, 196
  examples: functional spec, 197
  examples: test case, 199
  examples: user assistance, 196
  user questions, 192

  functional specifications and, 219
  how do I do it, 194
  how do we design it, 221
  how do we know it passed, 239
  how do we test it, 238
  procedures and, 194
  what can go wrong, 221
  what is it, 193, 219, 237
  why did it do that, 195
  why does persona need this to work, 237
  why does persona want, 220
  why should I care, 192
documentation, testing, 227
documenting defects, 232
dominant colors, 156
downtime, developer, 141
due dates, 77

E

early adopters, 134
early majority adopters, 135
earnings, retained, 66
education, in résumés, 98
email, 80
emotions, perceptions and, 188
emphasis
  all caps, 152
  bold, 151
  color and, 156
  combination, 151
  font attributes for, 151–152
  italics, 151
  underline, 151
empty stage space, 168
engineering and ethics, 101, 110
engineering project flow, 113
  business requirements, 116
  delivery, 124
  development and testing, 120
  functional specifications, 117
  reality check meeting, 123

specifications, 117
start, 114
technical specifications, 117
testing and, 230
enterprise level products, 3
environment, job search, 92
epics
    generating, 117
    in agile environments, 211
equations, 157
equitable vs. fair, teams and, 85
error messages, 195
    planning in functional specifications, 221
estimating projects, 123
ethics, 101
    beginning of, 101
    being who you say you are, 109
    engineering, 110
    family of origin and, 101
        example of, 104
    first-generation college students and, 102
    functional specifications and, 221
    mitigating product failure, 222
    personal decisions and, 105
    present and future, 106
    preventing product failure, 221
    testing documents and, 227
    use of AI, 55
    who you say you are, 106
    who you show you are, 107
        example of, 108
    your life story and, 106
experience
    describing in a résumé, 96
    in a cover letter, 92
    learning through, 178
explanations, 183–184
    cognitively constructed, 191
    complex vs. simple, 183
    creating with metaphors/analogies/similes, 246
    figurative language and, 245
    figure and ground in, 189
    grouping in, 190–191
    humans vs. animals, 184
    need to know why, 183
    perception speed and, 188
    perceptual, 185
    signal-to-noise ratio in, 188–189
    user questions, 192
        designing and, 196
        examples: functional spec, 197
        examples: test case, 199
        examples: user assistance, 196
        how do I do it, 194
        procedures and, 194
        what is it, 193
        why did it do that, 195
        why should I care, 192
    using metaphors/analogies/similes, 245

F

factory
    delivery, 126
    project managers, 122
    specifications, 119
failure, graceful, 222
failure, product
    mitigating, 222
    preventing, 221
fair vs. equitable, teams and, 85
familiarity, 246
family of origin, 101
    example of, 104
fast perception, 188
feature creep, 211
features
    compared to functions, 215
    deconstruction, 217
        functional specifications, 217
        test cases, 233
    personas and, 208, 213
    scenarios and, 213
    solving problems using, 215

specification elements
    action/result tables, 223
    business and functional requirements, 224
    description, 219
    how do we design it, 221
    priority, 219
    what can go wrong, 221
    what is it, 219
test case elements
    action/result tables, 241
    description, 236
    how do we know it passed, 239
    how do we test it, 238
    materials list, 239
    measurable/observable, 239
    naming/numbering, 235
    pass/fail metric, 243
    prerequisite tests, 240
    what is it, 237
figurative language, 245
    difficulty, 246
    familiarity, 246
    figurativeness, 246
    personas and, 247
    vs. examples, 247
figure and ground, 189–190
    role in organizing, 189
file naming conventions, 76
finance context, 129
finishing presentations, 170
first job résumés, 92
first-generation college students and ethics, 102
fixed costs, 64–65
flow, 146
    graphs/charts and, 157
flow of a project, 113
    business requirements, 116
    delivery, 124
    development and testing, 120
    functional specifications, 117
    reality check meeting, 123
    specifications, 117

start, 114
technical specifications, 117
testing and, 230
flow of money
    COGS and, 62
    communication and, 67
    overview, 61
    pitching ideas
        business cases, 137
        finance context, 129
        historical/technical context, 131
    revenue, 62
fonts
    attributes/emphasis for slides/UI, 151–152
    importance of using standard, 152
    serif vs. sans serif, 152
    size for slides, 164
    size for slides/UI, 151
    types of for slides/UI, 152
formatting graphs and charts, 159
formatting text, 151
    font attributes/emphasis, 151
    font serif vs. sans serif, 152
    font size, 151
four user questions, 192
    designing and, 196
    examples: functional spec, 197
    examples: test case, 199
    examples: user assistance, 196
    how do I do it, 194
    how do we design it, 221
    how do we know it passed, 239
    how do we test it, 238
    procedures and, 194
    what can go wrong, 221
    what is it, 193, 219, 237
    why did it do that, 195
    why does persona need this to work, 237
    why does persona want, 220
    why should I care, 192
functional specifications, 117, 120, 215
    deconstructing functions/features, 217

defects caused by unclear, 229
elements of, 219
    action/result tables, 223
    business and functional requirements, 224–225
    description, 219
    how do we design it, 221
    priority, 219
    what can go wrong, 221
    what is it, 219
    why does persona want, 220
function vs. feature, 215
    problem-solving, 215
structure of, 216–217
user questions, examples, 197
functioning in the workplace ecosystem, 69
large companies, 70
onsite work, 72
remote work, 72
    home office setup, 73
    professional behavior, 74
    standards and processes, 75
small companies, 70
work from home, 72
    home office setup, 73
    professional behavior, 74
    standards and processes, 75
functions
compared to features, 215
deconstruction, 217
    functional specifications, 217
    test cases, 233
personas and, 208, 213
scenarios and, 213
solving problems using, 215
specification elements
    action/result tables, 223
    business and functional requirements, 224
    description, 219
    how do we design it, 221
    priority, 219
    what can go wrong, 221
    what is it, 219

test case elements
    action/result tables, 241
    description, 236
    how do we know it passed, 239
    how do we test it, 238
    materials list, 239
    measurable/observable, 239
    naming/numbering, 235
    pass/fail metric, 243
    prerequisite tests, 240
    what is it, 237
funding, asking for, 137–143
future ethics, 106
future tense, 12
    uncertain timing, 12

G

gender, engineering and, 5
general & administrative costs (GA), 64–65
general availability, 124
generating epics, 117
generative AI, 51
    cautions, 56
    ethical use of, 55
    prompts, 53
    use cases, 55
gestalt, visual, 187
    emotions and, 188
    figure and ground, 189
    grouping, 190–191
    perception, 188–189
    perception speed, 188
gestures, in presentations, 167
goals, setting, 87
    SMART method, 87
gold master, 124
graceful failure, 222
grammar
    checker, 50
    in graphs and charts, 160
    in UI, 160

graphics in procedures, 34–36
graphs, 157–159
    bar charts, 158
    formatting, 159
    grammar and spelling in, 160
    line, 158
    pie charts, 159
    selecting the correct type, 159
gray space, 186
    in written documents, 187
gross profit, 63
    industry range, 63
gross profit margin, 63
ground, 189
group chat, 82
grouped procedures, 43
grouping
    in perceptual design, 190
    in perceptual explanations, 191
    in procedures, 37
guidelines
    communication, 7–25
    cover letter, 90
    gray space and, 187
    white space and, 187

H

habits, 180
    effect of standards on, 180
    impact of input modes on, 205
    learning and, 180
hand gestures, in presentations, 167
headers vs headings, 20
headings, 20–21
    black space and, 186
    design and, 21
    every paragraph (no), 21
    frequency of, 20
    online documents and, 20
    scanning and, 20
headings vs headers, 20

headlines in written documents, 187
hierarchy of testing documents, 232
historical context, 131
holding questions to end, 166
home office setup, 73
how do I do it, 194
    functional specifications and, 221
    test case and, 238
        manual vs. automated, 238
how humans think about the world, 171
how you behave and ethics, 107, 109
    example of, 108
human resources (HR), 86
    business requirements, 117
    communicating with, 8
humans
    attentive behavior, 174
    brains and, 173
    cocktail party effect, 176
    cognition of, 171–181
    color perception and, 173
    explanations and, 183–184
    impact of culture and biology on, 171–172
    learning, 177
        cognitive load, 180
        experience, 178
        habits, 180
        interference, 180
        schemas, 178
    metaphors/analogies/similes vs. examples, 247
    physical world and, 171
    preattentive behavior, 173
        habits and, 180
        standards and, 175
    redundant signals and, 177
    sensory adaptation, 177
    systems of information and, 171–172
    vision and, 173, 185
        space on page, 186
hypothetical archetype, 206

I

ideas, pitching, 129–143
    business cases, 137
images in procedures, 34–36
imperative voice, 10, 241
impostor syndrome, 102
information design, 146, 185
innovation, 132
    disruptive vs. sustaining, 133
    technology adoption curve and, 132
        schemas, 179
innovators, 134
input modes, 202–206
    aural/auditory, 203
    improving communication with, 203
    kinesthetic, 204
    learning theory and, 205
    reading, 204
    visual, 203
integration testing, 233, 239
    IT role in, 126
interfaces
    animation, 150
    color, 153
    consistency, 150
    contrast, 155
    designing for, 145
    font attributes/emphasis on, 151
    font size on, 151
    grammar and spelling in, 160
    graphs/charts and, 157
    maximum information and, 149–150
    mood, 156
    progressive disclosure, 149
    serif vs. sans serif, 152
    short text, 149
    structure/emphasis, 156
    text formatting on, 151
interference, learning, 180
intern résumés, 92
intuitive design, 185

IT
    delivery, 126
    professional services, 122
    project managers, 122
    specifications, 119
italic text, 151

J

job experience, 96
    metrics in, 97
job hunting, 98
    job shops, 99
    staffing agencies, 99
job listings, importance of following, 89
job searches and other resources, 89
jokes, in presentations, 165

K

kinesthetic input mode/VARK, 202, 204
knowledge base, 216

L

laggards, 136
large companies, 70
large language models (LLM), 51, 56
late majority adopters, 135
latinates, avoiding, 14
layering information in procedures, 42
leadership, 84
    teams and, 85
        dysfunction, 85
        problems, 85
learning theory, 177
    cognitive load, 180
    experience, 178
    habits, 180
    input modes and, 205
    interference, 180
    schemas, 178
Lile, Samantha, 159

line graphs, 158
listening to complete questions, 169
localizers
    delivery, 127
    development/testing, 122
low vision, 153

M

managers, writing for, 58
manual test cases, 238, 241
marketing
    delivery, 125
    specifications, 118
materials list, test cases and, 239
materials/procurement, specifications, 119
maximum bullet points on slides, 149
maximum length
    paragraphs, 18
    sections, 20
    sentences, 16
measurable, test cases, 239
meetings, 78
    Zoom fatigue in, 79
mental models, 178
mental processes
    attentive processes, 174
    cocktail party effect, 176
    combination, 174
    complex vs. simple explanations and, 183
    designing with user questions, 196
    explanations and, 183–184
    learning, 177
        cognitive load and, 180
        experience and, 178
        habits and, 180
        interference and, 180
        schemas and, 178
    need to know why, 183
        cognitive explanations, 191
        humans vs. animals, 184
        perceptual explanations, 185

preattentive processes, 173
    habits and, 180
    standards and, 175
sensory adaptation, 177
    redundant signals and, 177
user questions, 192
    how do I do it, 194
    procedures and, 194
    what is it, 193
    why did it do that, 195
    why should I care, 192
metaphors, 245
    compared to examples, 247
    difficulty, 246
    familiarity, 246
    figurativeness, 246
    personas and, 247
metric, 97
microphones, in presentations, 167
minimum viable product, 3, 117, 213
money, 59
mood, using color to set, 156
Moore, Geoffrey, 132
moving, in presentations, 167
multi-modal communication, 203

N

naming test cases, 235
need to know why, 183
    cognitive explanations, 191
        designing with user questions, 196
        examples: functional spec, 197
        examples: test case, 199
        examples: user assistance, 196
        how do I do it, 194
        user questions and, 192
        what is it, 193
        why did it do that, 195
        why should I care, 192
    humans vs. animals, 184
    perceptual explanations, 185

nested procedures, 41, 43
net profit, 65–66
Nielsen, Jakob, 201
noise-to-signal ratio, in perceptual communication, 188–189
numbered steps
    cognitive load and, 28
    maximum, 28
    procedures and, 43
numbering
    test cases, 235
    test documents, 233
numbers
    in graphs and charts, 157
    in résumés, 97

O

objectives, in résumés, 93–95
observable, test cases, 239
online help, 216
onsite companies, 72
operating environments, scenarios and, 210
operational health, 63
organizing
    figure and ground, 189
    grouping, 190
outline slides, 147
    vs. slide outline, 148
overall project flow, 113
overhead, 64–65
overview procedures, 38–39
O'Keefe, Sarah, x

P

page, 185
    visual space on, 186
    white space on, 186
paragraphs
    5-sentence limit, 18
    bulleted lists, 23
    example, 24–25

in written documents
    gray space and, 187
    white space and, 187
length of, 18
maximum length, 18
supporting sentences, 23
parts of speech, 9
    predicate, 9
    subject, 9
    verb, 9
pass/fail metrics in test cases, 243
passive voice, 10–11, 195
past tense, 11
peers
    testing presentations with, 162
    writing for, 57
people (see personas or scenarios)
perception
    active/fast/preattentive, 188–189
    attentive behavior, 174
    attentive processes, 174
    brains and, 173
    cocktail party effect, 176
    cognitive explanations, 191
    cognitive load, 180
    combination preattentive/attentive, 174
    complex vs. simple explanations, 183
    complex vs. simple explanations and, 183
    designing with user questions, 196
    emotions and, 188
    experience, 178
    explanations, 183–184
    explanations and, 183–184
    figure and ground, 189
    grouping, 190–191
    habits, 180
    human, 173, 185
    interference, 180
    learning, 177
    metaphors/analogies/similes, 246
    need to know why, 183
        cognitive explanations, 191

humans vs. animals, 184
    perceptual explanations, 185
preattentive behavior, 173, 180
preattentive processes, 173
    habits and, 180
    standards and, 175
redundant signals and, 177
schemas, 178
sensory adaptation, 177
    redundant signals and, 177
space on page, 186
standards and, 175
user questions, 192
vs. animals, 184
white space on page, 186
perceptual explanations, 185
    designing for
        figure and ground, 189
        grouping, 191
        perception speed, 188
    figure and ground in, 189
periods at the end of sentences, 9
personal decisions and ethics, 105
personal preferences and product development, 201
personas, 206–208
    context and scenarios, 210
    creating, 207
    defects caused by unclear, 229
    demographics/behavior, 207
    functional specifications and, 215
        action/result tables, 223
        why does persona want, 216, 220
    input modes, 202
        improving communication with, 203
    metaphors/analogies/similes, 247
    scenarios and, 209–212
    schemas and, 179
    test cases and, 237
    user assistance and, 216
    user-centered development and, 201, 213
    using, 208, 213
    wrong, impact of, 212

photographs in procedures, 35
phrases and words to remove, 16
pictures in procedures, 34
pie charts, 159
pitching ideas, 129
    business cases, 137
    finance context, 129
    historical/technical context, 131
    personal context, 132
    technology adoption curve and, 136
plain language, 9
planning, 87
    SMART method, 87
posture, in presentations, 167
pragmatists, 135
preattentive processes, 173
    figure and ground, 189
    grouping, 190
    grouping and, 191
    habits and, 180
    not conscious, 174
    perception speed and, 188
    standards and, 175
predicate, importance of, 9
preparing for presentations, 162
    day of, 164
prerequisite tests, 240
present tense, 11
presentations, 98, 145–160
    (see also slides)
    day of, 164
    delivering, 161–170
        day of, 164
        script reading, 162
        test run, 162
    demonstrating applications, 164
    designing slides for, 145
        clutter, 146
        flow, 146
        projected/emitted light, 147
        reflected light, 147
        simplicity, 146

empty stage space and, 168
finishing up, 170
hand gestures, 167
jokes in, 165
moving around, 167
politics in, 169
posture, 167
preparation, 162
    day of, 164
questions and answers, 168
    holding to end, 166
    listening to all of, 169
    rank and, 169
    restating, 169
    via chat, 168
    "I don't know", 168
reading slides, 168
script reading, 162
structure of, 148
    conclusion slide, 148
    outline slide, 147
    question slide, 148
    title slide, 147
test run, 162
time of, 165
upspeak, 167
using microphones in, 167
using your body in, 166–167
    hand gestures, 167
    moving around, 167
    posture, 167
    whole room, 166
    Z sweep, 166
using your voice in, 167
    empty stage space, 168
    reading slides, 168
    verbal tics, 168
verbal tics, 168
virtual, 167
voice, 167
whole room, 166
Z sweep, 166

prior art, 5, 130–131
priority, functional specifications and, 219
pro-rated costs, 64
problems
    solving with functions/features, 215
    solving with products, 131
        personal preferences and, 201–202
        personas/scenarios and, 206–212
        wrong personas/scenarios and, 212
problems in teams, 85
procedures, 27–48
    analogy to code, 27
    before you start, 30
    bold text in, 33
    chunking, 43
    combining overview and task-specific, 41
    common errors, 37
    complex, 42
    definition of, 28
    emphasis in, 34
    end user, 32
    example, 32–33
    goal of, 29
    graphics in, 34–36
    grouped/nested/chunked, 43
    grouping in, 37
    grouping steps, 44–45
    name of task, 29
    notes/cautions/warnings, 30
    numbered steps in, 31
    overview in, 29
    overview vs. task-specific, 38
    paragraphs in, 31
    programming readers, 27
    results of, 30
    structure of, 29
    subtasks, 46
    suitably vague, 38
    task paths in, 36
    task pause points, 46
    type
        combination, 41

complex, 42
  overview, 39
  task-specific, 40
user questions and, 194
process, test cases and, 238
processes and standards, 75
  deadlines/due dates, 77
  file naming conventions, 76
product failure
  mitigating, 222
  preventing, 221
product managers
  specifications, 118
product testing, 227–243
  costs of, 229
  deconstructing test cases, 233
  rationale, 228
  risk analysis and, 230
  when to start, 230
product-centric writing, 15
products
  personas and, 208, 213
  scenarios and, 213
products solve problems, 131
  personas/scenarios and, 201, 206–212
  wrong personas/scenarios and, 212
professional behavior, 74
  chat, 81
    group, 82
    presentations and, 164
    privacy and, 82
  email, 80
    presentations and, 164
    To/CC/BCC, 80
  meetings, 78
    Zoom fatigue, 79
  standards and processes, 75
    deadlines/due dates, 77
    file naming conventions, 76
  ticketing systems, 83
  work computers, 84
professional credibility, 75

professional résumés, 92
professional services
  delivery, 126
professors, writing for, 57
profit, 59–60
  business writing and, 58
  communication of, 67
  flow of money and, 61
  gross, 63
  net, 65–66
  pitching ideas, 129
  reinvestment of, 66
profit margin, 65
programming methodologies
  agile, 4
  waterfall, 4
progressive disclosure, 149
project estimation, 123
project flow, 113, 127
  business requirements, 116
    CEO, 116
    CFO, 116
    CMO, 116
    CTO, 116
    human resources, 117
    VP Sales, 116
  delivery, 124
    build master, 125
    developers, 125
    factory, 126
    IT, 126
    localizers, 127
    marketing, 125
    product managers, 126
    professional services, 126
    project managers, 125
    QA, 126
    sales, 125
    shipping, 127
    support, 126
    tech writers, 126
    trainers, 126

development/testing, 120
  architect, 121
  developers, 121
  factory, 122
  IT, 122
  localizers, 122
  product managers, 121
  professional services, 122
  project managers, 121
  QA/testers, 121
  support, 123
  tech writers/UI/UX, 122
  trainers, 122
  UI/UX, 121
functional specifications, 117
reality check meeting, 123
specifications, 117
  architect, 118
  developers, 118
  factory, 119
  IT, 119
  marketing, 118
  materials/procurement, 119
  product managers, 118
  project managers, 118
  QA/testers, 118
  sales, 118
  tech writers/UI/UX, 119
start, 114–115
technical specifications, 117
testing and, 230
project management, specifications and, 118
project start, 114
projected/emitted light, 147
projects, 98
prompts for AI, 53
proofreading
  graphs and charts, 160
  UI, 160
publications, 98

Q

QA/testers
  delivery, 126
  project managers, 121
  specifications, 118
quality assurance (QA), 227
  costs of, 229
  risk analysis and, 230
  when to start, 230
Quality Assurance Institute, 228
questions
  four user, 192
    designing and, 196
    examples: functional spec, 197
    examples: test case, 199
    examples: user assistance, 196
    functional specifications and, 219
    how do I do it, 194
    procedures and, 194
    what is it, 193
    why did it do that, 195
    why should I care, 192
  handling in presentations, 168
    rank and, 169
    via chat, 168
    "I don't know", 168
  holding to end, 166
  listening to all of, 169
  restating, 169
questions slide, 148

R

range of gross profit, 63
rank, handling questions in order of, 169
reading level, 14–19
  average in the US, ix, 14
  dumbing down and, 14
  non-native English and, 19
reading, input mode/VARK and, 202, 204
reality check, 123–124
  using priorities in the, 219

recessive colors, 156
recipes, as a model for test cases, 234–235
recipient information, business letter, 90
recurring revenue, 27
redundant signals, 153, 177
reflected light, 147
reinvestment, 66
remote companies, 72
    home office setup, 73
    professional behavior, 74
    standards and processes, 75
repetition
    clarity and, 13
    in writing, 13
    using it or this, 13
    value of, 13
reporting defects, 232
requirements
    business and functional, 224
    defects caused by unclear, 229
    implied in scenarios, 211
    testing, 230
researching personas, 207
resources for job searches, 89
restating questions, 169
résumés, 89
    awards/projects/presentations/publications, 98
    cover letters and, 90
        addresses, 90
        hiring manager name, 91
    education in, 98
    first vs. later job, 92
    importance of following job listings, 89
    job experience in, 96
    job search environment, 92
    job searches and other resources, 89
    metrics in, 97
    objective in, 93
    past tense, 96
    skills in, 95
    special skills in, 95
    structure of, 93

where to send, 98
work history in, 96
    metrics, 97
writing guidelines, 96
retained earnings, 66
return on investment (ROI), 77, 114, 138–139
    business example, 139
    personal example, 138
revenue, 62
    recurring, 27
risk
    analyzing, 230–231
    calculating, 231
    user comfort with, 134, 136
ROI (see return on investment (ROI))
room/space, using all of in presentations, 166

S

sales
    delivery, 125
    product managers, 126
    project managers, 125
    specifications, 118
salutation, business letter, 91
sans-serif fonts, 152
scenarios, 209–212
    agile environments and, 211
    example, 209
    implied requirements and, 211
    input modes, 202
        improving communication with, 203
    operating environments in, 210
    providing context for personas, 210
    schemas and, 179
    user-centered development and, 201, 213
    using, 211, 213
    wrong, impact of, 212
schemas
    cognitive load and, 179
    impact of input modes on, 205
    learning and, 178

personas/scenarios and, 179
   shifting, 179
   standards and, 179
   technology adoption curve and, 179
   users and, 179
scoping projects, 123
screen captures in procedures, 35
script reading and presentations, 162
scripts, using in test cases, 238, 241, 243
second person, 12–13
section headings, 21
sections
   5-paragraph limit, 20
   headings and, 20
   length of, 20
   maximum length, 20
senses, human, 173, 185
   brains and, 173
sensory adaptation, 177
   redundant signals and, 177
sentences, 9–10
   20-word limit, 16
   future tense, 12
   in procedures, 31
   in written documents, 187
   length of, 16
   maximum length, 16
   present tense, 11
   second person, 12
   subject+verb+predicate construction, 9, 16
   supporting, 23
   topic, 21
   words and phrases to remove, 16
serif fonts, 152
similes, 245–247
   difficulty, 246
   familiarity, 246
   figurativeness, 246
   personas and, 247
   vs. examples, 247
simple explanations, 183
simple risk analysis, 231

simplicity, 146
   graphs/charts and, 157
skeptics, 136
skills, highlighting in résumés, 95
skills, in résumés, 95
slide outline vs outline slide, 148
slides
   animation on, 149–150
   bullet points on, 149–150
      maximum, 149
   color and, 153
      contrast, 155
      mood, 156
      structure/emphasis, 156
   conclusion slide, 148
   consistency and, 150
   designing, 145
      clutter, 146
      flow, 146
      interfaces and, 145, 149–150
      projected/emitted light, 147
      reflected light, 147
      simplicity, 146
      UI/UX and, 145, 149–150
   font attributes/emphasis for, 151
   font size for, 151
   font types for, 152
   formatting text for, 151–153
      attributes/emphasis, 151
      font size, 151
      serif vs. sans serif, 152
   graphs/charts and, 157
   outline slide, 147
      vs. slide outline, 148
   questions slide, 148
   reading to audience, 168
   space on page and, 187
   timing, 149
   title slide, 147
small companies, 70
SMART method, 87
smoke tests, 125

social media
    cleaning up for a job search, 90
    on work computers, 84
solving problems, 131
    personal preferences and, 201–202
    personas/scenarios and, 206–212
    wrong personas/scenarios and, 212
speaking in presentations, 167
specifications, 117
    functional, 215
        action/result tables, 223
        business and functional requirements, 224
        deconstructing functions/features, 217
        description of features/functions, 219
        elements of, 219
        function vs. feature, 215
        priority of features/functions, 219
        structure of, 216–217
    technical, 217
spelling
    checker, 50
    in graphs and charts, 160
    in UI, 160
standards
    cognitive load and, 175
    effect on habits, 180
    schemas and, 179
standards and processes, 75
    chat, 81
        group, 82
        presentations and, 164
        privacy and, 82
    deadlines/due dates, 77
    email, 80
        presentations and, 164
        To/CC/BCC, 80
    file naming conventions, 76
    meetings, 78
        Zoom fatigue, 79
    ticketing systems, 83
    work computers, 84
STEM-adjacent, 4

steps, numbered, 28
stickiness, 27
stimulus
    sensory adaptation, 177
        redundant signals and, 177
stimulus/response table, 198, 233
stories, in a cover letter, 92
structure
    business letter, 90
        address, 90
        body of letter, 91
        close of letter, 91
        date, 91
        salutation, 91
    functional specifications, 216–217
        deconstructing functions/features, 217
    in slide/UI design
        color and, 156
    presentations, 148
        conclusion slide, 148
        outline slide, 147
        question slide, 148
        title slide, 147
    résumé, 93
        awards/projects/presentations/publications, 98
        education, 98
        metrics in work history, 97
        objective, 93
        skills, 95
        work history, 96
    test cases, 235
        recipe model, 234–235
subject+verb+predicate construction, 16
support, development/testing, 123
supporting sentences, 23
sustaining innovation, 133

T

task paths, 36
task pause points, 46
task-specific procedures, 38, 40–41

teams, problems/dysfunction in, 85
tech writers
    delivery, 126
    shipping, 127
    specifications, 119
    support, 126
technical communication, 10–11
    tools, 49
technical context, 131
technical debt, 3, 124
technical specifications, 117, 120, 217
    defects caused by unclear, 229
technology adoption curve, 132, 136
    business cases and, 136
    chasm, 134
    early adopters, 134
    early majority/pragmatists, 135
    innovators, 134
    late majority/conservatives, 135
    pitching an idea and, 136
    skeptics/laggards, 136
    use in a business case, 138
tense (past, present, future), 11–12
terminology, 3
    agile environment, 4
    attentive processes, 174
    black space, 186
    clutter, 146
    cocktail party effect, 176
    experience, 178
    feature creep, 211
    figure, 189
    flow, 146
    gray space, 186
    ground, 189
    grouping, 190
    habits, 180
    interference, 180
    minimum viable product, 3
    page, 185
    preattentive processes, 173
    prior art, 5

projected/emitted light, 147
reflected light, 147
schemas, 178
sensory adaptation, 177
simplicity, 146
STEM-adjacent, 4
technical debt, 3
waterfall environment, 4
white space, 186
terminology managers, 49, 51
test cases
    deconstructing functions/features, 233
    documenting, 227
    elements of, 235
        action/result tables, 241
        description, 236
        how do we know it passed, 239
        how do we test it, 238
        materials list, 239
        measurable/observable, 239
        naming/numbering, 235
        pass/fail metric, 243
        prerequisite tests, 240
        what is it, 237
        why does persona need this to work, 237
    hierarchy, 232
    numbering, 233
    structure of, 235
        recipe model, 234–235
    user questions, 199
test suites
    hierarchy, 232
    numbering, 233
testing, 120, 227–243
    costs of, 229
    ethics and, 227
    in agile and waterfall environments, 227
    in startup environments, 228
    integration, 233, 239
    presentations and, 162
    rationale, 228
    regulated industries, 228

risk analysis and, 230
unit, 233, 239
when to start, 230
testing phase, 124
text
    formatting for slides/UI, 151–153
        attributes, 151
        font size, 151
        serif vs. sans serif, 152
    input mode/VARK and, 204
ticketing systems, 83
timing, slide, 149
title slide, 147
tone, in colors, 155
tools, 49
    AI, 51
        cautions, 56
        prompts, 53
        use cases, 55
    chat, 81
        group, 82
        presentations and, 164
    email, 80
        presentations and, 164
        To/CC/BCC, 80
    generative AI, 51
        prompts, 53
    grammar/spelling check, 50
    terminology managers, 49, 51
    ticketing systems, 83
    work computers, 84
tooltips, 193
topic sentences, 21
    place in paragraph, 21
    using two short for, 22
    using 'and', 22
trainers
    delivery, 126
    project managers, 122
troubleshooting, 195
    planning in functional specifications, 221
Tufte, Edward, 159

types of procedures
    combination, 41
    complex, 42
    overview, 39
    task-specific, 40
Tyson, Neil DeGrasse, 184

U

UI/UX
    action/result tables and, 223
    designing for, 145
        animation, 150
        color, 153
        consistency, 150
        contrast, 155
        font size on, 151
        graphs/charts, 157
        maximum information and, 149–150
        mood, 156
        progressive disclosure, 149
        redundant signals, 153
        serif vs. sans serif, 152
        short text, 149
        structure/emphasis, 156
        text formatting on, 151
    development/testing, 121
    functional specifications and, 223
    grammar and spelling in, 160
    project managers, 122
    space on page and, 187
    user assistance and, 216
underlined text, 151
unexplained events, explanations and, 183
unit tests, 233
upspeak, 167
usability, user-centered development and, 201
user assistance, 216
    user questions, examples, 196
user stories, 201, 211
    (see also personas and scenarios)
user-centric writing, 14–15

users, 201
    (*see also* personas or scenarios)
    personas and, 206–208
    questions, 192
        designing and, 196
        examples: functional spec, 197
        examples: test case, 199
        examples: user assistance, 196
        how do I do it, 194
        how do we design it, 221
        how do we know it passed, 239
        how do we test it, 238
        procedures and, 194
        what can go wrong, 221
        what is it, 193, 219, 237
        why did it do that, 195
        why does persona need this to work, 237
        why does persona want, 220
        why should I care, 192
    scenarios and, 209–212
    schemas and, 179
    user-centered development and, 201, 213
        wrong personas/scenarios, 212
    when not to use the term, 12
    when to use the term, 13

V

value, 59
    profit and, 59–60
variable costs, 62
VARK, 202
    improving communication with, 203
    learning theory and, 205
verbal tics, 168
virtual presentations, 167
vision, 173, 185
    creating meaning, 188–189
        figure and ground, 189
        grouping, 190–191
    emotions and, 188
    space on page and, 186

visual
    input mode/VARK and, 202–203
visual gestalt, 187
    emotions and, 188
    figure and ground, 189
    grouping, 190–191
    perception, 188–189
    perception speed, 188
visual space, 187
visual, input mode/VARK and, 202–203
visuals in procedures, 34–36
voice
    active, 10–11
    active and action verbs, 10
    imperative, 10, 241
    passive, 10–11
    using in presentations, 167
VP Sales
    business requirements, 116
    project start, 114

W

waterfall environments
    functional specifications in, 215
    testing in, 123, 227, 230
WebAIM contrast checker, 156
what is it, 193
    functional specifications and, 219
    test cases and, 237
what's in it for me, 192
white space, 186
    in written documents, 187
who you say you are and ethics, 106
who you show you are and ethics, 107, 109
    example of, 108
why did it do that, 195
    functional specifications, 221
    test case and, 239
why should I care, 192
    functional specifications and, 220
    test cases and, 237

words and phrases to remove, 16
work computers, 84
work from home (WFH)
    home office setup, 73
    professional behavior, 74
    standards and processes, 75
work history, 96
    metrics in, 97
work product, 4
workplace ecosystem, 69–88
    contributor, 69
    functioning in, 69
    goal-setting/planning, 87
        SMART method, 87
    large companies, 70
    leadership, 84
        teams, 85
    onsite, 72
    remote, 72
        home office setup, 73
        professional behavior, 74
        standards and processes, 75
    size, effect on company culture, 70
    small companies, 70
    work from home (WFH), 72
        professional behavior, 74
        standards and processes, 75
    work-from-home (WFH)
        home office setup, 73
writing
    academic, 7, 57–58
    business, 7, 57
        profit and, 58
    repetition in, 13
    technical, 57
    user-focused, 14
writing for managers, 58
writing for peers/professors, 57
written communication
    clarity of
        business goals, 8
        functional specifications, 217

technical specifications, 217
            user assistance, 216
        importance of, 8
written documents, space on page in, 187

Y

your life story and ethics, 106

Z

Z sweep and presentations, 166
Zoom fatigue, 79

Colophon

About Sharon Burton

Sharon Burton consults as a technical content strategist and teaches part time at the University of California, Riverside, in the Bourns College of Engineering. Leveraging her graduate cultural anthropology education, her career has been built around making content and content development easier for both businesses and content consumers. In that time, she's also taught upwards of 8,000 people in corporate and university level courses.

In her spare time, she knits, sews, cooks, grows food, bakes bread, and rides her bike, can be found on the back of a motorcycle, and goes to the gym. Three or more nights a week, Sharon teaches baby engineers to communicate at the University of California, Riverside, as part of the engineering program.

About Bonni Graham Gonzalez

Bonni Graham Gonzalez teaches part time at the University of California, Riverside, in the Bourns College of Engineering. Bonni's career has focused around storytelling, including designing and writing manuals that actually explain how to use products and creating award-winning marketing materials that both educate and sell. When not marketing or thinking about marketing, Bonni crochets, beads, cooks, plays with her dog, and goes on road trips with her husband.

About XML Press

XML Press specializes in publications for technical communicators, content strategists, marketing communicators, and managers. We focus on concise, practical publications concerning content strategy, management, and XML technologies.

www.ingramcontent.com/pod-product-compliance
Lightning Source LLC
Chambersburg PA
CBHW061344210326
41598CB00035B/5875